EMPOWERED BY
THE DEVIL

THE RISE AND FALL OF ADOLPH HITLER

David H. Sorenson

New Harbor Press

New Harbor Press
1601 Mt Rushmore Rd, Ste 3288
Rapid City, SD 57701
www.newharborpress.com

Ordering Information:
Quantity sales. Special discounts are available on quantity purchases by corporations, associations, and others. For details, contact the "Special Sales Department" at the address above.

Empowered by the Devil/Sorenson—1st ed.

ISBN 978-1-63357-409-0

First edition: 10 9 8 7 6 5 4 3 2 1

All Scripture quotations are taken from the Holy Bible, King James Version (Public Domain).

Contents

Foreword

By *vocation*, the author of this book has been a pastor and theologian for over 50 years. He holds several graduate degrees in theologically oriented fields. By *avocation*, he is an historian of World War II, who has read over 100 books on the subject.

After 50 years of being a pastor, he understands human nature. This background of working with people and their problems over the decades has given unique insights into human nature that secular scholars or historians might not have. We all are of one race, the human race. An understanding of human nature thus offers clues into the persona and personality of Adolf Hitler.

Because of his over 50 years of dealing with people and their issues, the author has insights into why people do what they do. This background, along with substantial research on World War II, coupled with a clear understanding of Scripture, enables insight into why and how a man like Adolf Hitler wound up doing what he did.

The author further has a scriptural perspective of history. God has a plan for the nations, particularly His chosen nation, Israel. There is a clear blueprint within Scripture of how end-time events will unfold. Though the events of World War II and Adolf Hitler did not fulfill those prophecies per se, nevertheless, those events have set the stage for the impending end of the age. God certainly was aware of the events of World War II. Indeed, He quite apparently allowed it all to happen for His divine purposes. Though not the purview of this book, a case can be made that God providentially assisted the Allies at crucial points of the war to ensure its outcome (e.g., against all odds, the American victory at the Battle of Midway or winter coming early in Russia in 1941). And, in God's Providence, the catastrophic events of World War II have set the stage for the coming culmination of the age.

Introduction

One of the most forceful personalities of the 20th century was Adolf Hitler. Few loved him outside pre-war Nazi Germany. Most of the world hated and feared him. Nevertheless, he was without question one of the most influential leaders in the 20th century. As a result of the war he started, the world at the end of 1945 was dramatically different than in August of 1939.

Hitler started what undoubtedly was the greatest secular event in human history—World War II. It was the greatest war in world history involving nations from every continent, Antarctica excepted. World War II certainly involved the war in the Pacific and eastern Asia. However, that will not be the focus of this book. We will rather focus on what took place in Europe.

World War II was the worst war in history with more people dying through its course than at any other time in history. An estimate of 70-85 million people perished. It was the first war where more civilians than soldiers perished—an estimated 50 million civilians. Adolf Hitler in considerable measure was responsible for that horrendous number. It is the considered opinion of this author that Adolf Hitler was a *prototype* of the coming Antichrist, clearly foretold in Scripture, and was empowered by the Devil.

The man responsible for the majority of the conflict was Hitler. Understanding the times and culture in which he grew up is insightful. Likewise, understanding his personal character and how he rose to power will give further understanding of the man.

This book will provide an overview of World War II, focusing particularly upon Hitler and his decisions, of major geo-political issues which shaped the war, and of the major battles of the European Theater of Operations (ETO). We will also take note of major personalities and weapons of the war.

We will not endeavor to overwhelm the reader with tedious details of individual battles, or with an exhaustive history of the war. Rather, we will give highlights and insights of what caused World War II, why Hitler did what he did, his connection to the occult, and focusing upon the major blunders he made which spelled irrevocable doom for Nazi Germany.

CHAPTER 1

Germany in the 1920s

In the 1920s, Germany was a defeated nation. It had been one of the instigators of World War I, but had lost the Great War, particularly after the United States entered as an ally of Britain and France.

Germany was not invaded or occupied by her foes at that war's end. Neither were her cities or industries bombed or destroyed. Rather, an armistice was signed on November 11, 1918. The armistice was essentially a cease fire. The Armistice did not end the First World War officially, but it was an agreement which halted the fighting on the Western Front while the terms of the permanent peace were hammered out.

On June 28, 1919, the Treaty of Versailles was signed at the Hall of Mirrors in the Palace of Versailles, Paris, France. It officially ended World War I. Germany and the three principal European Allied powers, Britain, France and Italy were the signatories. Ironically, the United States did not sign.

The Treaty of Versailles held Germany responsible for starting the war and imposed harsh penalties in terms of loss of territory, massive reparation payments, disarmament, and demilitarization. Far from the "peace without victory" that U.S. President Woodrow Wilson had outlined in his famous Fourteen Points in early 1918, the Treaty of Versailles humiliated Germany while failing to resolve the underlying issues that had led to war in the first place. Economic distress and resentment of the treaty within Germany helped fuel the ultranationalist sentiment that led to the rise of Adolf Hitler and his Nazi Party, as well as the coming of another World War just two decades later.

As a result, Germany was an economic basket case in the 1920s. Hyper-inflation ravaged the nation's economy. In 1923, when the battered and heavily indebted country was struggling to recover from the disaster of the First World War, cash became very near-ly worthless. Such inflation began during World War I when the German government printed unbacked currency and borrowed money to finance its dream of conquering Europe. Germany could not import the goods it needed for survival. Accordingly, daily life became difficult for many. Hyperinflation created a sit-uation whereby prices rose almost by the hour. People were paid twice a day and often had to take piles of money to the store in wheelbarrows.

Germany was thus humiliated by the Treaty of Versailles and that humiliation ran deep and wide throughout the nation. The coun-try was embittered and blamed France in particular for much of it.

As a result, Germany chafed at the onerous reparations from the Great War. The Treaty of Versailles (signed in 1919) and the 1921 London Schedule of Payments required Germany to pay 132 billion gold marks ($33 billion USD) in reparations to cover civilian damage caused during the war. These burdens simmered in Germany for years. They particularly left an indelible mark on a German-Austrian war veteran by the name of Adolf Hitler. In fact, he made it his life-long goal to overturn what he and many others perceived as the grave injustices foisted upon Germany by the Treaty of Versailles. That national humiliation and embitterment was the seed plot from whence the Nazi Party would arise.

In addition to the onerous burdens imposed, many in Germany held the view that the Jews were the real instigators of Germany's troubles. Antisemitism was widespread and widely accepted. One man in particular would seize upon that prevailing notion of anti-Jewish hatred and that man was Adolf Hitler. He blamed Germany's woes not only on France and the Treaty of Versailles, but on European Jewry and the quickly rising communist movement, many of whom were in fact Jews.

Germany in the 1920s and 30s was thus yearning for a leader who would restore Germany to her former prominence, restore the German economy, rebuild the German military, cast off the onerous provisions of Versailles, and recompense France in particular. World War II, which emerged in September of 1939, in reality was a re-ignition of the smoldering hostilities which the Treaty of Versailles never properly addressed. World War II in fact was the inevitable culmination of how World War I had been put on hold 20 years earlier.

CHAPTER 2

Hitler's Background and Ascension to Power

Adolf Hitler was born April 20, 1889 near Linz, Austria, and was the fourth of six children of Alois Hitler and his third wife, Klara Pölzl. Hitler's personal lawyer, Hans Frank, would later reveal in his memoir the following story about Hitler's family tree.

In 1930, Hitler, after a query from his nephew, directed Frank to investigate his heritage. Alois' mother, Maria Anna (Adolf's grandmother), had been employed as a housekeeper by a Jewish family in Graz, Austria; and the family's 19-year-old son Leopold Frankenberger had been carrying on with her. As a result she wound up pregnant with Alois. If that is in fact true, then Alois' son, Adolf, would be a quarter Jewish. No definitive records remain, though there is little question that Alois Hitler was an illegitimate child.

Such situations were not unusual in Austria of that era. But a prominent Jewish family surely would seek to keep such a birth quiet. Alois Hitler's mother, Maria, was Catholic but his baptism record did not list the name of his father. However, what is telling is that the Frankenberger family paid child support to Alois' mother until Alois was 14 years old. In his memoir, Frank recounted that when he presented his findings to Hitler, Adolf initially denied nothing, including the Frankenberger's support payments for his father. Adolf Hitler later would claim that his grandmother blamed Leopold Frankenberger as a means to extort child support from them. It should be noted, however, that the source of the *initial* account was the personal and rather thorough investigation by Hans Frank, Hitler's own lawyer in 1930, before Adolf Hitler came to power.

Meanwhile, Adolf Hitler later in life had the records of that community for that time destroyed which only adds more suspicion to the matter. (The Nazi Party initially defined a Jew as one having just one Jewish grandparent, though it would later conveniently be re-defined as one having three Jewish grandparents.)

The eight-year-old Adolf took singing lessons, sang in the Catholic Church choir, and even considered becoming a Catholic priest. When his younger brother Edmund died of the measles, Adolf changed from a confident, outgoing, conscientious student to a morose, detached boy who constantly fought with his father and teachers. Adolf's father had made a successful career as a bureaucrat in the local customs' office. He desired that Adolf follow in his footsteps. Adolf rather wished to attend a classical high school to study to be an artist. His father forced him to attend a regular high school. In rebellion against his father and that decision, Adolf would later write that he purposely did poorly so that his father would allow him to pursue his desires.

Though Austrian, Adolf developed a close affinity for nearby Germany. After the death of his father in 1903, Adolf's behavior and school performance improved. But he graduated from high school only after having failed the final exam and had to retake it.

His education was unremarkable and he never went any further academically than a high school diploma.

In 1907 young Hitler moved to Vienna to study fine art. That same year his mother died and he ran out of money, subsisting by living in homeless shelters. He eked out a living doing casual labor and by painting and selling water colors of local landmarks. While living in Vienna, Hitler also was exposed to virulent anti-Semitism. He would later embrace it after World War I as a product of the "stab-in-the-back" explanation for the catastrophe.

When World War I broke out in 1914, Hitler was living in Munich. At the age of 25, he enlisted in the Bavarian Army where he served as a dispatch runner on the western front rising to the rank of corporal. He was wounded at the Battle of the Somme and was decorated for bravery, receiving the Iron Cross, Second Class. On a recommendation by Lieutenant Hugo Gutmann, Hitler's Jewish superior, he received the Iron Cross, First Class on August 4, 1918, a decoration rarely awarded to one of Hitler's lowly corporal rank. On October 15, 1918, he was temporarily blinded by mustard gas and was hospitalized.

Hitler's wartime experience reinforced his German patriotism. He was aghast at Germany's capitulation on November 11, 1918. That bitterness over how the war ended for Germany began to shape his ideology. Like many Germans, he believed the "stab-in-the-back" idea which claimed the German army, "undefeated in the field" had been "stabbed in the back" on the home front by civilian leaders: Jews, Marxists, and those who signed the armistice that ended the war.

The Treaty of Versailles demanded that Germany give up territories and demilitarize the Rhineland region of Germany abutting the French border, along with heavy reparation payments to France and England. Hitler, along with many in Germany saw this as a deep humiliation, especially the paragraph of the treaty which declared Germany as responsible for the war. These deep-seated grievances would become the impetus for Hitler's political career.

After the war, Hitler, with no education higher than barely a high school diploma and no career prospects, therefore remained in the army. He was assigned to be an intelligence agent of the army which was suspicious of the nascent German Workers' Party (GWP)—the precursor of the coming Nazi Party. Hitler's army superiors ordered him to join the party as an under-cover agent, thinking any party with the term *workers* in it might be a communist front group. But Hitler liked what he found in the GWP. There, he was also introduced to the Thule Society which was a German occult organization, tracing its roots to the so-called Aryan race. Meanwhile, the German Workers' Party changed its name to the *Nationalsozialistische Deutsche Arbeiterpartei* or National Socialist German Workers' Party (NSDAP). Its contracted form was "Nazi."

After leaving the army in 1920, Hitler went to work for the Nazi Party full time. Their focus then was to eradicate Marxism as well as undermine the weak sitting government, known as the Weimar Republic of Germany. Even at this early point in his political career, Hitler had become a public speaker, gaining a reputation for his angry speeches against the Treaty of Versailles, and particularly against Jews and Marxists. Nevertheless, there was bitter infighting within the party. However, on July 29, 1921, he was appointed and given absolute power as party chairman.

During the 1920s, Hitler was given books on the occult from his association with the Thule Society. Hitler evidently read them. He underlined passages extensively, including one which read, "He who does not have the demonic seed within himself will never give birth to a magical (i.e., occult) world" [Ernest Schertel, *Magic: History, Theory and Practice.*] Dietrich Eckhart was an occultist and a member of the Thule Society. He was called the *spiritual* founder of National Socialism to whom Hitler dedicated his book *Mein Kampf.* Eckhart was credited with indoctrinating the young Hitler in the occult. Eckhart would later write, "Follow Hitler! He will dance, but it is I who have called the tune. We have given him the 'means of communication' with *Them*" [Herbert Brennan, *Occult Reich*]. The *them* in its context referred to demonic spirits. Another infamous occultist by the name of Alice Bailey would write, "That Adolf Hitler was possessed by the Dark Force" [Alice Bailey, *The Externalization of the Hierarchy*].

Alfons Heck was a former member of the Hitler Youth, who went on to become a fanatical adherent to the Nazi Party during the Third Reich. He recalled those early years and Hitler's speeches.

> "We erupted into a frenzy of nationalistic pride that bordered on hysteria. For minutes on end, we shouted at the top of our lungs, with tears streaming down our faces: Sieg Heil, Sieg Heil, Sieg Heil! From that moment on, I belonged to Adolf Hitler body and soul."

Sieg Heil is a German phrase which means "Hail to Victory." It would become the basic verbal salute of the Nazi Party until its dying day. Both the Nazi arm salute as well as the party symbol, the swastika, found their origins in occult societies.

Adolf Lanz was a notable occultist in Austria. In 1932, he wrote a letter to an occult order, the year before Hitler came to power. There, he wrote, "Hitler is one of our pupils. . . . He will one day be victorious and develop a movement that makes the world tremble." That prophecy quite evidently came to pass. It was Lanz in 1907 who first ran up the Swastika flag over an occult temple in Vienna. CBS reporter in Germany, William Shirer reported that it was Hitler who personally adopted the Swastika as the symbol of the Nazi party.

At about this time, Hitler began to manifest an unusual ability to manipulate and mesmerize his audiences. As will be further documented shortly, he already was involved with the occult. This author believes he became empowered by the Devil himself. He could virtually hypnotize crowds with his passionate rhetoric. In the greater scope of history, men of God have been empowered by the Spirit of God to accomplish supernatural feats for divine purposes. The Devil likewise has such abilities. He found in Hitler, an empty vessel waiting to be filled with the power of darkness, especially as it pertained to attacking the Jewish race. Bluntly, Hitler's rise to power was assisted by the Devil himself. In him, Satan saw a useful instrument to wreak vengeance upon the Jewish people.

In 1923, Hitler and associates attempted a coup against the Weimar Republic, the sitting government of Germany. Along with the help of World War I General Erich Ludendorff, Hitler staged the infamous Beer Hall Putsch in Munich. (A *putsch* was a German word for a violent attempt to overthrow the government.) On November 8, 1923, Hitler along with Ludendorff stormed an assembly of 3,000 men at a major beer hall in Munich. He fired a pistol into the ceiling and announced the national revolution had begun and he declared the formation of a new government with Ludendorff. They then proceeded to occupy a local police station. The next day they marched to the Bavarian War Ministry to overthrow the Bavarian government, but police dispersed them. Sixteen Nazi Party members and four police officers were killed in the failed coup. Hitler was arrested three days later and tried for high treason. He was sentenced to five years in prison, but was pardoned by the Bavarian government and released less than a year later.

However, Hitler did not waste his time in prison. Rather, he dictated a book which came to be known as *Mein Kampf* which basically means, "my struggle." As noted earlier, the book was dedicated to Dietrich Eckhart, a member of the occult Thule Society. In it, Hitler laid out his plans for transforming German society based upon race and eliminating the Jews. He likened international Jewry to germs which had infected Europe. His simple solution was to exterminate them.

Additionally in *Mein Kampf*, Hitler outlined his plans to expand Germany to provide more *lebensraum* (i.e., living room or living space). Specifically, he detailed an invasion of Russia and the Ukraine in particular. His plan was to conquer for Germany the rich agricultural lands, especially of the Ukraine region. He also detailed his disdain for the Slavic nationalities, calling them *untermencsh*—inferior or subhuman: gypsies, Poles, Serbs, Slavs, and Russians. Ironically, when Hitler did in fact invade the Soviet Union and its Ukraine region, he was fulfilling in detail what he had written about 18 years earlier. Evidently, the Russians never paid any attention to the book. Over time, *Mein Kampf* went on to sell over a million copies and made Hitler independently wealthy.

However, in 1924, Hitler was still an impoverished misfit. After failing to overturn the government through a putsch (i.e., coup d'état), he and the party determined to try and assume power in Germany through conventional political means. From 1924 to 1933, the fledgling Nazi Party ran candidates for the national legislature called the Reichstag. In 1924, they received 6.5% of the vote. By 1933, they had received about 44% of the vote which was more than any other German political party. But that is not all of the story. Through those years, they bullied voters, sending goon squads of party storm troopers through cities intimidating voters. Hitler was in the forefront of the skullduggery and violence. Of course, he continued his demonically-inspired speeches and tirades. His themes were always the same: overturning their own government which he alleged had stabbed Germany in the back at the end of World War I, vindicate Germany of the humiliation of Versailles; and above all, blame the Jews and communists for all of Germany's troubles.

The Great Depression which had begun in New York in 1929 with the crash of the New York Stock Exchange spread to Germany, adding to its already considerable woes. But this proved to be a boon for the Nazis. They capitalized on the nation's economic distress to leverage more political power. The worse things got economically, the more the Party prospered.

In 1932, Hitler ran against war hero Paul von Hindenburg for president of Germany. Hitler came in second with more than 35% of the vote, but in so doing became a force to be reckoned with in German national politics. Through the convoluted politics of the day, Hitler was appointed Chancellor of Germany later in 1932. The Nazis gained seats in the national cabinet. Hitler insisted and was allowed to take control of the police of Germany.

In 1933, a Dutch communist set fire to the Reichstag (i.e., capitol) building. As a result and again at Hitler's urging, an edict was issued which essentially suspended civil rights and detention without trial known as the "Reichstag Fire Edict." Later that year, amidst turbulent political events with Nazi thugs threatening other arriving members of the parliament, two major events took place. (1) A piece of legislation known as the Enabling Act was passed which gave Hitler's cabinet the power to enact laws without the consent of the Reichstag. (2) Hitler became the *de facto* dictator of Germany—its Fuhrer. He now was more or less unfettered in implementing his evil policies.

In June of 1934, Hitler boasted to a British correspondent,

> "At the risk of appearing to talk nonsense I tell you that the National Socialist movement will go on for 1,000 years! . . . Don't forget how people laughed at me 15 years ago when I declared that one day I would govern Germany. They laugh now, just as foolishly, when I declare that I shall remain in power."

Hitler continued to consolidate his power. In 1933 the Nazi Party was declared the only legal political party in Germany. He then moved to purge rivals and threats from within his own movement, personally directing the murder of Ernst Rohm, once a close compatriot in the Nazi movement. In August 1934, Hindenburg died and Hitler emerged as the head of State of Germany. He appointed himself as head of the military. He now had complete control of Germany.

CHAPTER 3

Hitler's Persona

Was Hitler a mad man? This author has read accounts of people of that era and even today referring to him as such. There is little evidence of that. Though obsessed with a twisted ideology which certainly influenced his decisions as a leader, Hitler ruled with a clear mind. Only in the last weeks of his life when his physical health was deteriorating and when he was living like an animal in his subterranean bunker in Berlin did he lose touch with reality. He was not a madman nor insane, though his perverse and evil policies might have made him seem that way.

Was Hitler stupid? To the contrary, he led Germany out of the Great Depression and rebuilt its economy. Though lacking in any higher education, he had an uncanny ability to out maneuver and manipulate the leaders of numerous nations. Though he had only achieved the rank of corporal in World War I, yet he was the primary planner and architect of Germany's initial military onslaughts between 1938 and 1941. He never had had any formal training in military strategy or campaigns. His generals were often reluctant and hesitant in the early military adventures into Poland, the Low Countries, and France. Yet, Hitler often proved to be right tactically and even strategically.

Hitler was *Time Magazine's* man of the year for 1938, as well as appearing on the cover of *Time* on multiple other occasions.

He secretly re-armed Germany and built it into the most powerful military in the world of that day. He personally planned and executed the German invasion, conquest, and occupation of most of Europe. Hitler was not stupid, rather, he was an evil mastermind. But perhaps his successes and accomplishments were empowered from one greater than him.

CHAPTER 4

Satanic Influence

As noted earlier, this author's background is an historian by avocation, but a trained, professional theologian by vocation with several advanced graduate degrees. By way of emphasis, I therefore will contend once again that Hitler's meteoric rise to power was through satanic assistance and empowerment.

The Devil is a real personality. The term *devil* is more of a title; Satan is one of his names. He was once one of three archangels in heaven above, perhaps chief of the three. But he sought to usurp the position of the Almighty and led a rebellion against Him. He is an unseen, powerful spirit-being who has long been at war with the Almighty, having rebelled against Him in the mists of antiquity. The Scripture is clear that God Almighty therefore threw Satan out of heaven, along with his cohorts, now known as demons. Their abode and domain in this age is planet earth where they seek to thwart the purposes of God and deceive the human race.

The popular misconception is that the Devil presently lives in hell, shoveling coal. However, as one author (Hal Lindsey) entitled a book some years ago, Satan is alive and well on planet earth. The Scriptures clearly teach there is a definitive command structure in the Devil's unseen empire, not unlike that of a military chain of command. He seeks to influence humanity not only on an individual level, but particularly through leaders in high places. He thus seeks to manipulate human events. In America for example, it is not uncommon for an elected official to go to Washington with a clearly stated agenda, only to completely change his or her views after some time. Unclean spirits of the Devil's empire likely have sought to influence them to accomplish Satan's purposes.

There are several degrees of unseen spiritual operations of the Devil upon people. (1) He (or his lieutenants called demons) seek to *influence* people through any number of means whether by manipulating education, religion, media, or social contacts, *et al.* (2) The Devil may *oppress* people by bringing the power of darkness upon them, leaving them mentally ill, depressed, despondent, and even suicidal. (3) In some limited cases, the Devil or his agents seeks to *possess* an individual whereby they endeavor to take control of that person. It is the professional judgment of this author that through contact with the occult in his younger-adult years, Adolf Hitler was possessed by the Devil to accomplish and empower him for his purposes.

A primary gateway into the realm of the Devil is the occult. A textbook definition of the occult might be, "a category of supernatural beliefs and practices which generally fall outside the scope of religion and science, encompassing such phenomena involving otherworldly agency as mysticism, spirituality, and magic." That surely is a euphemistic and verbose definition. More simply, the occult is interacting with the spirit world. It is in fact the anteroom of the realm of the Devil. To be sure, not a few practitioners of the occult are fakes, seeking attention for themselves. More often than not, they endeavor to increase their bank accounts through phony fortune-telling schemes. But the fact is, there is a real occult world and as noted above it is the gateway to the realm of Satan. What is significant is that Hitler and many of the early Nazis were involved in the occult.

In addition to Adolf Hitler, another arch villain of history was possessed by the Devil. In the hours before the crucifixion of Jesus Christ, the Devil entered (i.e., possessed) Judas Iscariot. John 13:27 records, *"And after the sop Satan entered into him."* The *him* referenced here is Judas Iscariot. Judas became the traitor of the ages in betraying Jesus Christ. The point here, however, is that the Devil can and does "enter" or possess individuals. He possessed Judas Iscariot; it is the contention of this author that he possessed Adolf Hitler as well.

In the greater overview of Scripture, God selected one nation as His chosen people long ago. That people to this day is the nation of Israel, more commonly known as the Jews. Because the nation of Israel has historically been God's chosen people, they have been a target of Satan. History is replete with a long record of the persecution of the Jews. The Devil is ultimately behind it all. Scripture in fact says that Satan has long sought to "persecute" the Jews (Revelation 12:13). The Devil is the ultimate anti-Semite. In fact, all antisemitism, at its root, goes back to Satan.

It is not the purview of this book to present a detailed description of God, the Devil, or of Israel. But in the end, Israel will someday be exalted as a nation when Christ returns and Satan will eventually be condemned to hell—forever. He knows that, but seeks to do as much damage as he can in the meantime. Among other things, he endlessly has persecuted the Jewish people if for no other reason than to spite God and perhaps to try and thwart God's plan.

In Hitler's hatred of the Jews and his obsession with anti-semitism, the Devil found a willing vessel to try and eradicate God's chosen people. Satan thus helped and empowered Hitler to that end.

Early in his political career, Hitler became involved with the occult—the Devil's domain. He received an ability to absolutely mesmerize and virtually hypnotize hundreds of thousands of Germans who came to Nuremberg and elsewhere to hear him speak (i.e., preach). Ironically, the term *mesmerize* derives from a Dr. Franz Anton Mesmer, a nineteenth century European occultist, who sought to entrance or hypnotize people through the spirit world. Hitler indeed "mesmerized" hundreds of thousands of listeners in his frenetic rallies. There clearly was a supernatural power flowing through him as he spoke, but the power was not from on High.

Ironically, Hitler was not a particularly gifted public speaker. His voice was harsh. He was repetitive. He was verbose. But his speeches usually began modestly and built in intensity. As he hit his stride at the rostrum, waves of raw power flooded from him capturing his audiences. Not a few observers have described his oratory as hypnotic. In one account, a man from England was in the audience at one of Hitler's rallies. He did not speak or understand German. But as Hitler rose into a frenzy, the Briton became so aroused that he found himself giving the Nazi salute and shrieking "Heil Hitler."

Moreover, Hitler's powers of persuasion were not limited to just mass rallies. Numerous high-level German military leaders after the war spoke or wrote of when in Hitler's presence a strange power seemed to sway them, even against their will. Numbers of them spoke of a strange power in his eyes. They often were opposed to Hitler's policies, both politically and militarily, but when meeting with the Fuhrer face to face, they were basically mesmerized to acquiesce to his wishes. They often left their meetings with Hitler holding a view they had opposed prior to their session with him.

Thule Society

The Thule Society was a German occult group, active particularly in the 1920s. The group was founded by Karl Haushofer and a World War I veteran by the name Walter Nauhaus. Before the war, Nauhaus belonged to another occult order known as the Order of Teutons. One of the main focuses of the Thule Society was the pursuit of the origins of the Aryan race. Hitler joined the Thule Society in 1920. Heinrich Himmler was also a ground-floor member. By some accounts, Himmler himself was an occult medium. Curiously, Dr. Theodore Morell, the quack doctor who virtually ruined Hitler's health in his final years, was also a member of the Thule Society. In fact, it was Morell who introduced Hitler to the Society. After the Nazi Party had formed, it later purchased the German newspaper called the *Munchener Beobachter* (the *People's Observer*) from the Thule Society. It originally had been the house organ of the Society.

The Thule Society would force their initiates to sign a "declaration of faith" concerning their Germanic heritage. Specifically, the signer swore to the best of his knowledge and belief there was no Jewish or colored blood that flowed in either his or in his wife's veins, and that among their ancestors are no members of the Jews or colored races.

Himmler took from the Thule Society much of its occult ideas and used them to help start the Ahnenerbe in 1935. The Ahnenerbe, which means ancestral heritage, was a branch of the SS. Their goal, was to make Germany as Aryan as possible by making Nordic connections wherever possible and claiming they were the purest of all races through their own versions of science and research. Disciplines such as *astrology, parapsychology and clairvoyance* were a significant part of the Thule Society and its adherents. The Schutzstaffel, also known as the SS, headed by Himmler used occult Nordic runes on their uniforms as a symbol of their believed origins.

The Power of Darkness

The Nazi rallies were frequently under cover of darkness with torch light parades and ceremonies with spotlights forming pillars of light into the night sky. The Devil's domain has always been the power of darkness, not only figuratively but literally. Not surprisingly, most occult activities take place at night under cover of darkness. So did Nazi rallies.

Demonic Charisma

On numerous episodes in Hitler's career, there are records of his outbursts, tantrums, and unhinged behavior. Though these are usually attributed to him being unbalanced psychologically— and that may have been the case, they are also symptomatic of demonic possession. In light of the greater context of Hitler's known history, his obsessive hatred of the Jews, and his angry outbursts; these are all consistent with one empowered by the Devil. A staff officer at OKW (the German high command), Colonel Ulrich Maiziere, after the war spoke of Hitler's "demonic charisma." He went on to say that Hitler had an "indescribable, demonic effect on other people, whom only very, very few people were able to resist."

Madame Blavatsky

Helena Petrovna Blavatsky was a Russian occultist and author who cofounded the Theosophical Society in 1875. She in fact was one of the chief proponents of the occult in the latter portion of the 19th century. The Theosophical Society she founded, and continues to this day, is a mainstream occult organization. Again, the occult is the vestibule to the realm of the Devil. In the late eighteen hundreds, Blavatsky authored at least five books widely circulated in occult circles. One report out of Russia after the fall of the Soviet Union is that the Russian soldiers who entered Hitler's bunker in Berlin after his death, found a copy of one of Blavatsky's books on his night stand beside his bed. Though that account cannot be documented, there is little question that Hitler was in fact involved with the occult through the Thule Society.

It is the conclusion of this author that Adolf Hitler was influenced and empowered by the power of the Devil. That is the simple and logical conclusion of how this backwards, ill-educated, misfit of society could rise meteorically from obscurity to become the most powerful man in the world of that day. He very likely was a *prototype* of the long-foretold coming Anti-Christ who likewise will be empowered and promoted by the Devil himself.

CHAPTER 5

The Personal Character of Hitler

A part from his Spartan lifestyle and modest appearance, Hitler's personal character was utterly corrupt. This is consistent with one used by the Devil. Ironically, Hitler did not drink or smoke.

Throughout his career, he showed himself to be a *pathological liar* and devious; telling the appeasement-seeking Chamberlain, for example, he was for peace when he in fact was planning for war. He likewise assured other European leaders of his peaceful intentions, all the while planning their invasion. He signed a non-aggression pact with Stalin in 1939, knowing full well that he intended to break his word not long after.

He was a *thief*. He, in cahoots with Hermann Goering, systematically looted art and art work all across Nazi occupied Europe. Hitler stole the wealth and possessions particularly of the Jews of Europe.

He was a *murderer*. Not only were his policies toward the Jews murderous and for any others he disliked, Hitler also personally ordered and directed the murder of his rival and former close associate Ernst Rohm. Hitler was utterly cruel in directing his forces to slaughter those whom he considered sub-humans or other undesirables to the Nazis.

He was *immoral*. Heinrich Hoffman, an early member of the Nazi Party, was Hitler's official photographer—for about 25 years. He knew Hitler well on a personal level. After the war, Hoffman was arrested for war crimes and was incarcerated at the Nuremberg jail awaiting trial. There, an American Navy lawyer was assigned to him whose name was Richard Heller. (Hoffman would be convicted and served four years in prison for war profiteering.) Hoffman however confided to his lawyer Lieutenant Heller revealing information about Hitler and his morals in the years before he became the Chancellor of Germany.

Among other things Hoffman ran a photography shop and from there he also published pornography. His models were girls who worked in second-rate bars—including an attractive seventeen-year-old blond named Eva Braun. In addition to posing nude for Hoffman, Eva was also sexually involved with him, a middle-aged, married man. Hoffman described Eva Braun as "coarse." From a professional pornographer, that was a revealing comment about her moral character.

Hoffman also observed that Hitler, as a middle-aged man, had an inordinate interest in young girls. In 1926 when Hitler was 37 years old, he began an affair with an underage 16-year-old girl named Mitzi Reiter which continued, on and off, for years. Another teenager was Hoffman's seventeen-year-old daughter, Henriette. Therefore to distract Hitler from his seventeen-year-old daughter, Hoffman introduced the forty-two-year-old Adolf to another seventeen-year-old, Eva. As the saying goes, the rest is history. Seventeen-year-old Eva Braun became Hitler's girlfriend and mistress. Though some have tried to portray Hitler's relationship with her as mostly platonic, the fact is they secretly slept together and lived together on and off for the next 14 years. Also of interest is that Henriette Hoffman went on to marry Baldur von Schirach who became the head of the Nazi Hitler Youth organization.

Meanwhile, there is a significant body of evidence that Hitler was also having a romantic affair with his eighteen-year-old niece, Geli Raubal, when he was 41 and 42 years old. In fact, they lived together for a period of time. She committed suicide September 18, 1931, under mysterious circumstances. But it was just prior to then that Hitler had taken up with Eva Braun. His niece likely became aware of that, provoking her suicide. In any event, there is clear insight into the morals of Hitler.

The last 14 years of his life, Hitler continued his ongoing affair with Eva Braun: his girlfriend, then mistress, and finally his wife for less than two days. This was all the while he assiduously worked to publicly portray himself as the paragon of personal morality and virtue to the German people. Only Hitler's closest inner circle was aware of his long-standing affair with Eva Braun.

Though Hitler himself lived a simple, even Spartan life style; as his wartime decisions developed, it is evident the man was covetous and greedy, ostensibly on behalf of German and its aggrandizement. In that day, Hitler was Germany and Germany was Hitler. As his successes and advances became pronounced, Hitler became proud and arrogant. The hubris he developed became one of the chief traits which led to his disastrous decisions, ensuring the defeat of Germany. Hitler was a deeply flawed man, apart from his coming monstrous policies

CHAPTER 6

Events Leading to World War II

U pon rising to power in Germany in 1933, Hitler immediately began to openly re-arm Germany. Actually, Germany had been secretly re-arming since shortly after World War I. German officers had secretly cooperated with the Russians in development of new weapons all through the 1920s and into the 1930s. But Hitler shifted the re-armament of Germany into high gear in 1933. In 1935, he publicly repudiated the Treaty of Versailles. A military draft system was inaugurated whereby the newly reconstituted Wehrmacht (i.e., the German army) began to grow rapidly. Hitler rejected the restrictions in the treaty upon the Kriegsmarine (i.e., the German Navy) and began to rapidly rebuild it. He likewise gave birth to the modern Luftwaffe (German air force). All the while, Britain and France, principal signatories to the Treaty of Versailles, did nothing. By the late 1930s, Nazi Germany was quickly becoming the most powerful nation on earth militarily. Though publicly proclaiming peace, Adolf Hitler planned to overrun Western Europe, especially wreaking vengeance on France. His ultimate goal was the establishment of a Nazi empire, reaching from the Atlantic to the Ural Mountains in Russia.

As Germany re-armed, German industry sprang back to life. The once basket-case economy of post-war Germany, further depressed by the Great Depression of the 1930s, suddenly became vibrant. Not only was Germany rising in power militarily, so was its economy. The once wide-spread unemployment of the German work force suddenly became fully employed. Times became good in Germany. Prosperity reigned. The average German gave Adolf Hitler the credit.

After the humiliation of Germany in the Treaty of Versailles, Hitler had a burning lust for vindication, power, and prominence in Europe. Though Hitler was the instigator of the coming war with all its atrocities, most ordinary Germans supported him in his decisions. As early as 1936, Hitler began to implement his long-nurtured dreams of building his German empire.

CHAPTER 7

Hitler's Conquest of Western Europe

From 1936 until the spring of 1941, Hitler began to systematically overrun and conquer Western Europe. Notice the counterclockwise sequence to nations overran by Germany.

The Re-militarization of the Rhineland

On March 7, 1936. Hitler marched into the Rhineland, a section of Germany between the Rhine River and the German border with France and Belgium to the west. It included the heavily industrialized region of the Ruhr Valley. When World War I ended, part of the Treaty of Versailles demanded the de-militarizing of that section of western Germany—the Rhineland. The de-militarization was intended to increase the security of these nations against future German aggression. This area of Germany was also important for coal, steel, and iron production with much of Germany's production thereof located there.

Sending his army into the Rhineland was a direct violation of the Treaty of Versailles as well as the Treaty of Locarno. The latter was ratified in 1925 and established the western borders of Germany with its neighbors. France, Germany, and Belgium pledged to treat those border as inviolable. As signatories of the agreement, Britain and Italy committed themselves to help repel any armed aggression across the frontier.

Though Hitler was the instigator of sending troops into the Rhineland, his own generals were reluctant—they feared provoking France. Furthermore, German generals knew at that point, Germany was in no position for another war with France. The German Chief of the General Staff warned Hitler that the army would be unable to successfully defend Germany against possible French retaliatory action. Hitler went so far as to assure his nervous generals that he would withdraw his forces if the French made a counter move. The operation was code named Winter Exercise.

Hitler had a special train carrying him and other high ranking officials taken to near the incursion area. He spent much of that night fretting over what would happen the next day. Far from the confidence and hubris of later German campaigns, even Hitler was nervous about what would happen. The German general staff considered Hitler's action suicidal.

Not long after dawn on March 7, 1936, nineteen German infantry battalions and a handful of planes entered the Rhineland. When German reconnaissance learned that French troops were assembling along the French-German border, German General Blomberg begged Hitler to withdraw. Hitler informed his generals that he would wait until the French actually reacted. It turned into an international game of chicken. Hitler called the French bluff. His first use of force went unchallenged.

When German troops marched into the city of Cologne in the Rhineland, cheering crowds formed to greet the soldiers, throwing flowers onto the Wehrmacht while German Catholic priests offered to bless the soldiers. All over Germany, there were wild celebrations as news reached the nation of the re-militarization of the Rhineland. The nation rejoiced. It was not until Germany defeated France in 1940 that the Nazi regime was again as wildly popular across Germany as it was in March of 1936. To try and legitimize his victory, Hitler called for a referendum on March 19, 1936 for a post facto approval. The Nazi party claimed that there was a 99% vote of "yes." Whether it was that high is open to debate, but there is no question that the German public overwhelmingly supported Hitler's unilateral action.

For a variety of reasons neither Britain nor France did anything. Though there was some huffing and puffing, Hitler had successfully called their bluff. France at that time, at least on paper, was more than adequate militarily to have crushed the Wehrmacht in the Rhineland. But France was tied in knots in its internal politics. Britain had no stomach to send troops to the continent after the bloodletting of World War I. Hitler's first military foray succeeded without the firing of a shot.

The Anschluss of Austria

On March 12, 1938 through bullying, intimidation, and threats, Hitler annexed Austria to Germany. Austria was Germany's immediate neighbor to the south. Moreover, Austria was a German speaking country which had had close ties to Germany over the centuries. In 1871 the Prussian-dominated German Empire (Second Reich) had excluded Austria. Over the ensuing 67 years, Austria had developed as a nation in its own right. However, there had been talk of annexation with Germany from time to time.

In 1938, Nazis within the Austrian government conspired to seize the government by force and unite it with Nazi Germany.

Upon learning of the conspiracy, the Austrian Chancellor, Kurt von Schuschnigg, traveled into Germany and met with Hitler, hoping to assure the independence of his Austria. Hitler treated him rudely, making him wait long hours before seeing him and then intimidating him with German military commanders present in the room. Hitler thereupon bullied him into naming several top Austrian Nazis to his cabinet.

On March 9, 1938, Chancellor Schuschnigg called for a national referendum to determine once and for all the question of Anschluss (i.e., annexation) to Germany. Furious over this, Hitler brought further pressure to bear upon Schuschnigg and forced him to resign on March 11, before the vote. Under pressure from Austrian Nazis, taking orders from Hitler, Schuschnigg pled with Austrian forces not to resist a German "advance" into the country.

The next day, March 12, German forces rolled across the border into Austria. As per the plea of the former chancellor, Austrian forces did not resist. Once the dust settled, Hitler himself was driven into Austria to a hero's welcome. Enthusiastic crowds greeted Hitler and his army. On March 13, the Anschluss (i.e., annexation) of Austria to Germany was proclaimed. As he had done in the Rhineland, Hitler then ordered a referendum in Austria after the fact. On April 10, 1938 in a vote controlled by the Nazis, 99.7% of the people allegedly voted in favor of the Anschluss. Austria therefore became a federal state of Germany. (After World War II was over, the Allies re-established Austria as an independent state.) Chancellor Schuschnigg meanwhile was imprisoned by Hitler until the end of World War II.

What remains remarkable is that Hitler overran a neighboring country without firing a shot. However, behind the scenes, there had been bullying, intimidation, and threats by a powerful nation to its weak neighbor. Moreover, lest there be any question about the matter, the German army had invaded Austria by force of arms. But whether by hook or by crook, Hitler was on a roll which in three years would find him the ruler of most of Europe.

Hitler Occupies Czechoslovakia

After annexing Austria in March of 1938, Hitler turned his attention toward Czechoslovakia. The western portion of Czechoslovakia, called the Sudetenland, was like a peninsula sticking into the side of southeastern Germany. Approximately three million German-speaking citizens lived there. Hitler directed German-speaking Nazis in Czechoslovakia to incite trouble which led to a crisis between the two nations. Hitler planned to annex the Sudetenland as he had Austria. The President of Czechoslovakia, Edward Benes, was not a happy camper. Apart from losing a portion of his country to a belligerent Germany, much of the Czechoslovakian national defenses as well as the large Czech defense industries were located in the Sudetenland, precisely to defend against German aggression. Should Germany gobble up the Sudetenland, Hitler would also acquire Czechoslovakia's basic national defense system.

In August of 1938, British Prime Minister Neville Chamberlain sent an emissary, Lord Runciman, to the Sudetenland to see if a settlement could be reached between Czechoslovakia and Germany. He returned to England empty handed.

On September 15, 1938, the British Prime Minister, Neville Chamberlain, himself flew to Germany and met with Hitler. They agreed on Czechoslovakia ceding the Sudetenland to Germany. The French Prime Minister Edouard Daladier also agreed upon the plan. Ironically, no representative of Czechoslovakia was invited to this meeting.

Seven days later, September 22, Chamberlain met again with Hitler to confirm the agreement. But Hitler now demanded not only annexation but the immediate German occupation of the Sudetenland. Hitler then made a major speech proclaiming that the Sudetenland was "the last territorial demand I have to make in Europe." He gave Czechoslovakia a deadline of September 28 to cede the Sudetenland to Germany or face immediate war.

At this point, Italian dictator Benito Mussolini suggested a conference be held in Munich of the major parties involved. Hitler, Daladier, and Chamberlain met and, agreeing to Mussolini's proposal, signed what came to be known as the Munich Agreement. The essence of the agreement was the immediate occupation of the Sudetenland by Germany. Once again, the Czechoslovakian government was not invited to the talks. Nevertheless, they agreed two days later to abide by the terms of the agreement.

On September 30, 1938, Chamberlain flew back to London. On the tarmac of London's Heston Aerodrome, as he debarked from his plane, he waved the Munich Agreement in the air and announced to the world that the Munich Agreement meant "Peace for our time." He ever since has carried the opprobrium of having appeased Hitler.

Hitler promptly invaded Czechoslovakia. On October 1, 1938, German troops occupied the Sudetenland: Hitler had gotten what he wanted without firing a shot. Six months later in March of 1939, a year almost to the day of the Austrian Anschluss, Hitler took over the rest of Czechoslovakia. He could add another notch to his gun. Moreover, he had lied through his teeth to the world that he had no other territorial ambitions. In less than a year, he devoured Poland, igniting World War II. But once again, and to all appearances, Hitler seemed to be a geo-political genius. He continued to proclaim to the world that he was a man of peace.

Though his successes had been achieved through bluffing, bullying, intimidation, and lying; it worked. Over the next several years, he would continue his string of unbroken success.

CHAPTER 8

Hitler Invades Poland

As Hitler and Nazi Germany increasingly became a threat to the peace and security of Europe, Britain and France became nervous. It was increasingly apparent that Hitler's next target was Poland. In March of 1939, France and Britain announced their guarantee of Polish independence. At that time German East Prussia was separated from Germany proper by a Polish corridor with Danzig as a free port city on the Baltic Sea. The Polish Corridor was Poland's only access to the sea. The corridor separating East Prussia from the rest of Germany was another result of the Treaty of Versailles which Hitler despised. In 1939, Hitler therefore demanded a right-of-way through the Polish Corridor to build an autobahn (i.e., super highway) and railroad between Berlin and Konigsberg in East Prussia. Poland refused. In August 1939, Hitler delivered an ultimatum to Poland—agree to his demands or face war. Ironically, Hitler confided to his generals that he was using the Danzig corridor issue as a pretext. His real goal was taking over Poland for further *lebensraum* (i.e., living space) for Germany.

Meanwhile, Hitler had negotiated a non-aggression pact with the Soviet Union in August of 1939. In it, both parties agreed not to attack the other. (Neither were ready for war with each other as yet.) However, secretly included in that pact was an agreement to carve up Poland between Germany and Russia. The pact also allowed Russia and Germany a secret protocol in which the independent countries of Finland, Estonia, Latvia, Lithuania, Poland and Romania were divided into spheres of interest of between Russia and Germany respectively.

On August 25, Britain signed the Polish-British Common Defense Pact. In it, both countries pledged to go to war on behalf of the other if either were attacked. Nevertheless, on September 1, 1939, Germany invaded Poland. Britain and France both issued ultimatums to Germany to withdraw from Poland or they would declare war on Germany. Hitler ignored them and on September 3, 1939, both Britain and France declared war on Germany. Hitler's foreign minister Von Ribbentrop had assured Hitler that Britain and France would not intervene just as they had sat on their hands when Germany re-militarized the Rhineland and overran Austria and Czechoslovakia. Hitler had not wanted war with Britain, but this time England and France called his bluff. Technically, World War II had begun.

The Gleiwitz incident

On the day before the Wehrmacht punched into Poland, an incident was staged under the auspices of the German SS and the Gestapo. (The Schutzstaffel—security staff—was the military branch of the Nazi Party and usually abbreviated as the SS.) On the night of August 31, a small group of German operatives dressed in Polish uniforms seized the German radio station at Gleiwitz which was next to the Polish border. They proceeded to broadcast a short anti-German message in Polish. Their intent was to make it seem that Polish forces had crossed the border into Germany and sabotaged a German radio station, thus giving Hitler a pretext to invade Poland in reprisal.

To make the incident seem more realistic, the Gestapo murdered a German farmer by the name of Honiok, known to sympathize with the Poles. He had been arrested the previous day by the Gestapo and dressed to look like a saboteur, then killed by lethal injection. The Gestapo thence fired gunshots into his dead body. He was left dead at the scene so that he appeared to have been killed while attacking the station. In addition, several prisoners from the nearby Dachau concentration camp were also murdered and left at the site making it appear that they had assisted Honiok. Then, the Germans made sure that the news media saw the scene and were led to believe that German authorities had driven off the attack, killing those left behind. It became one of several ugly incidents along the Polish border which the Nazis staged to blame Poland and use as justification for German action. Hitler cited the border incidents in a speech to the Reichstag that same day, with several of them called very serious and justifying his invasion of Poland.

The Invasion

The Wehrmacht attacked Poland the day after the Gleiwitz incident from the south, west and north. It was largely a one-sided battle. Germany implemented a new kind of warfare which came to be called *Blitzkrieg* or "lightning war." This was distinctly different than any military tactic to which Europe had been accustomed. During World War I, armies marched forward on foot, slowly and ponderously, slugging it out in planned, fixed-piece battles with the enemy, who likewise pounded out battle formations on the ground. Blitzkrieg was a method of offensive warfare designed to strike swift, focused blows at the enemy using mobile, maneuverable forces, especially utilizing modern tanks and close air support. The Germans sliced through the totally unprepared Polish army.

The Germans utilized their Panzer I and Panzer II light tanks. The Poles had a superior tank, but only 140 of them had been built. The Germans invaded with 2,000 tanks. The Poles had 445 obsolete planes in their air force. The Germans attacked with upwards of 2,000 planes, basically destroying the Polish air force on the ground the first day of the war. Many of the German attack planes were Junkers JU-87 dive bombers, more commonly known as Stukas. They were a single engine, low wing, all metal aircraft in distinction to archaic Polish, fabric-skinned bi-planes. Stukas were state of the art for the late 1930s. They were designed for close air support of German infantry and armor. The Stuka carried two .30 caliber machine guns and could carry up to 500 pounds of bombs. (Later models even carried a modest cannon under a wing.) They were designed for close air support and were essentially flying tanks, closely coordinated with German ground forces. One of their favorite tactics was dive bombing enemy targets and then returning to strafe other targets of opportunity. They had mounted on their undercarriage a siren which was activated during attacks. The wailing, screaming sirens were intended to strike terror into the enemy, which often they did.

The German blitzkrieg utilized massed, fast tank formations with motorized-mechanized infantry producing a rapid offensive, overwhelming enemy positions. Overhead were the screaming Stuka dive bombers blasting enemy formations just ahead of the advancing German armor and mechanized infantry. The tactics utterly overwhelmed the ossified, obsolete Polish army.

Roughly 1.5 million German soldiers crossed the border into Poland. Although the Polish army had one million soldiers, they were severely under-equipped with obsolete equipment at best. They even fielded horse cavalry against German armored divisions. The results were predictable. Moreover, Poland attempted to fight the Germans head-on, rather than falling back to more natural defensive positions. The Polish campaign was a debacle for Poland.

Meanwhile on September 17, the Russians invaded Poland from the east to seize those sections of eastern Poland that Hitler and Stalin had secretly agreed upon in their non-aggression pact. The war was over on October 6, 1939, though the primary fighting lasted for about three weeks. It had taken Germany less than a month to invade, defeat, and devour Poland. Germany lost 16,343 troops killed with 30,300 wounded. Poland on the other hand lost 66,000 troops killed, 133,700 wounded, and 694,000 captured. Or put another way, total German casualties were less than 47,000 while the Poles suffered over 827,000 casualties or soldiers captured. Poland ceased to exist.

Though Hitler surely had losses in this campaign, Germany lost only about 6% of what Poland suffered. Once again, Hitler was victorious. His armies had demolished Poland. His string of victories was growing larger. He had another notch in his gun. Accordingly, his hat size increased as well.

CHAPTER 9

The Phony War

After declaring war against Germany, England and France did not do much. For eight months into May of 1940, there was little action on the so-called western front. Because of this inactivity, this period came to be known as the "phony war." In Germany, it was called the "sitzkrieg" or sitting war. There was some action, but not much.

England immediately began to pour troops into France beginning on September 9, 1939. This buildup of troops in France continued until the debacle at Dunkirk. By then, England had landed 390,000 troops on French soil, along with thousands of heavy weapons and vehicles from tanks to trucks.

The French made a half-hearted attack in the Saar region of Germany, but quickly retreated upon facing German resistance. They basically decided to stay on the defensive and let the Germans bring the war to them. That they did eight months later. Ironically, if the French had put their full strength and heart into attacking Germany, they likely may have succeeded and ended the war then. Because the bulk of the German army was occupied in Poland, they had their hands full there. France at this time had one of the largest armies in the world, with 110 divisions available to attack Germany. But the French leadership was fossilized. German generals after the war confessed, that if France had pressed the issue, Germany was not prepared to deal with them. Germany had only 23 divisions available for defense on their border with France. German General Siegfried Westphal said that if the French had attacked in force in September 1939 the German army "could only have held out for one or two weeks." But France basically did nothing.

The Germans for their part conscripted more than one million new men into the Wehrmacht during this time. German armament industries used the time to further ramp up production of armaments and weapons. In the first few months of the war, Hitler hoped to persuade Britain to negotiate a peace settlement. He did not want a two-front war, though he had provoked it. Furthermore, he considered the British to be Aryan cousins inasmuch as England over the centuries had been settled by Saxons, Normans, and Vikings from Scandinavia. In Hitler's book, that made Anglo-Saxon England racially near of kin to his Aryan Germans. But because the British spurned his peace overtures, Hitler would within a year vent his wrath against the island nation.

Meanwhile, the British were openly discussing invading Norway to intercept shipments of iron ore from Sweden across northern Norway and thence by the Norwegian inner-coastal waterway to Germany. German ships could basically navigate behind the numerous coastal islands of Norway and avoid the open ocean and the Royal Navy. This prompted the Germans to take preemptive action and invade Norway and Denmark in April 1940 before the British could get their act together. Though there was some fighting in northern Norway between German and British units, the German prevailed and drove the British out.

Also during this time in Britain, Prime Minister Neville Chamberlain resigned in May of 1940. Even his own Conservative Party was disgusted with his appeasement of Hitler and the Munich Agreement, allowing Hitler to invade Czechoslovakia. Winston Churchill succeeded Chamberlain. He was a polar opposite of Chamberlain, unwilling to appease Hitler and determined to fight him.

The Germans began to ramp up attacks at sea against British ships, military and civilian. Both sides did some sporadic bombing of each other, though accomplishing little militarily. For eight months, it was a phony war between the Britain, France, and Germany. However, things would explode in May of 1940.

CHAPTER 10

The Invasion of Denmark and Norway

On April 9, 1940, Hitler simultaneously invaded Denmark and Norway—the same day. Denmark was Germany's immediate neighbor to the north. It held strategic value because it controlled the straits leading out of the Baltic Sea into the North Sea and the North Atlantic. Denmark shared a border with Germany and it was in the German's interest to have their northern border secured. Germany also saw Denmark's airfields as having strategic value for them—Germany would have additional bases from which to bomb England. Defensively, the German's intended to set up radar units along the Danish west coast for early warning of British bombing raids and other operational action.

Sweden had become the principal supplier of iron ore to Germany, feeding its blast furnaces to make steel for the German war machine. The Swedish iron ore mines were located in the far north of Sweden at a place called Kiruna. This region was near the borders of Norway, Sweden, Finland, and the Soviet Union, well above the Arctic Circle. Because of the high latitude, water-borne shipments of iron ore down the Baltic Sea were problematic during the long Nordic winter. The finger of the Baltic reaching up into that region was the Gulf of Bothnia. It froze solid during the long winter, making transshipment of iron that way impossible. The solution was the Swedes shipping their iron ore by railroad into nearby Norway to the Port of Narvik. From there, it was shipped by coastal freighters down the inner-coastal waterway of Norway and thence to Germany, largely avoiding the open sea. However, with the developing war with England, that route was at risk of attack by the Royal Air Force and Navy.

Germany also saw the numerous fjords of Norway as ideal places to base their submarines. From there they had direct access to the North Atlantic and would not have to traverse the North Sea, patrolled by the British.

As the Phony War languished, England openly discussed invading Norway. Their intent was to interdict the iron ore shipments out of Narvik, proceed inland, and perhaps even conveniently occupy the iron mines at Kiruna, Sweden. Moreover, control of the Norwegian coast would serve to tighten the blockade against Germany. The Germans were well aware of British intentions. However, Britain wrangled politically over what, when, and how much they should become involved in Norway.

The Germans were not so encumbered. In December of 1939, they began planning an attack on Norway. A high ranking Norwegian military officer and Nazi sympathizer by the name of Vidkun Quisling went to Germany and met with Hitler. Quisling would secretly work internally in Norway to pave the way for the Germans to occupy his country. His name would become synonymous with the word *traitor* ever since. The Germans developed an elaborate plan for the invasion of Norway called Operation Weserübung (i.e., "Exercise on the Weser"). Six divisions were planned for the invasion, with three going ashore immediately with the other three to follow. Other airborne paratroop battalions were assigned to seize airfields. The attack on Norway would coincide the same day with an invasion of Denmark.

Norway

A complex naval operation was planned. Instead of utilizing commercial cargo ships as troop transports, the Germans decided to transport most of their troops aboard warships. The latter were fast and could defend themselves. German ships began sailing on April 3 and all were coordinated to arrive at their assigned cities at the same time at 7 a.m., April 9, 1940. Every significant Norwegian coastal city from Oslo to Narvik was targeted. Meanwhile, Quisling had sent orders to these various ports to stand down and allow the Germans to dock. That the Germans did and immediately landed their troops unopposed. Every significant coastal city in Norway was attacked and occupied within 24 hours. Other towns were bombed at that same hour causing considerable damage.

For the most part, the Germans came ashore and quickly seized control of the various areas assigned to them. However, things did not go well for them near the capital city. Oslo, Norway, is located at the end of a long narrow fjord. At a particularly narrow strait called Drøbak Sound, the Norwegians had long before built a fortress with a powerful naval shore battery. Its two main guns were 11 inch caliber (280 mm) and were named Moses and Aaron. They were old, but still powerful. However, because of their age, the Germans did not consider them a threat.

The flagship of the German invasion was the heavy battle-cruiser Blucher and led a flotilla of German vessels. As the Blucher came abreast of the fortress, its main guns opened fire at point blank range devastating the Blucher. At the same time, the Norwegians fired torpedoes at close range which stuck below the waterline. The blazing ship sank immediately. Approximately 800 Germans were killed and another 1,400 survivors made it to land only to be captured by Norwegian forces. The other vessels in the German flotilla fled the scene. Many of the dead Germans were officers who were intended to direct the occupation by Germany along with a contingent of Gestapo agents.

The debacle for the Germans however, only slowed them and did not stop them. Other German forces were airlifted and landed at a nearby airport. Oslo was occupied later that night. However, the battle at Drøbak Sound had slowed the German timetable, enabling the King of Norway to board a train and escape to the north. He shortly thereafter flew to safety in England.

Meanwhile the British, aware of German action, landed troops at Narvik only to be driven off by the Wehrmacht. The British did manage to sink ten German destroyers near Narvik as well as several other German ships. Sporadic fighting continued for two months, but the issue was never in doubt. Germany had attacked, invaded, and occupied Norway. After the war, Quisling was arrested, tried, and executed for treason.

Denmark

Early on April 9, 1940 German forces also rolled into Denmark. At 4:00 a.m., the German ambassador to Denmark called the foreign minister of Denmark and demanded an immediate meeting with him. They met 20 minutes later and the German representative informed his Danish counterpart that German forces were already moving into Denmark, ostensibly to protect the country from Britain. There was a grain of truth to that. If British troops could land in Denmark, they would have a straight shot to Berlin to the south. Germany preempted that. Their forces had in fact already landed by ferry at 3:55 a.m.

The German ambassador demanded that Danish resistance cease immediately and that communication be made between Danish authorities and the Wehrmacht. If German demands were not met, the Luftwaffe would commence bombing Copenhagen. Additionally, German paratroops had already taken several Danish airports, as well as a strategic bridge and an old Danish fortress. German troops quickly landed at Copenhagen and captured the Danish garrison there. The Danish royal palace was attacked only to be temporarily turned back by the king's royal guard. This gave the king of Denmark, time to confer with his ministers and the head of the Danish army. German bombers appeared over the city and dropped leaflets urging Denmark to surrender.

At approximately 6:00 a.m., King Christian and the entire Danish government capitulated. In return, they requested retaining political independence in domestic matters. The Germans granted the latter, if only on paper. The invasion of Denmark had taken only several hours and was the shortest military campaign conducted by the Germans in World War II.

With Denmark in the bag and Norway in the process of being captured, Hitler had two more notches on his gun. The Norwegians and the British had managed to give Germany a bloody nose, in sinking the Blucher, two light cruisers, ten destroyers, and an assortment of several U-boats, transports, and other smaller warships. England lost one aircraft carrier, two cruisers, seven destroyers, one submarine, and several other assorted smaller vessels. But in the end, Germany prevailed. Hitler continued to have an unbroken winning streak, from the re-militarization of the Rhineland, to the Anschluss of Austria, to the invasion and occupation of Czechoslovakia, to the defeat of Poland, erasing it from the map. And now, Hitler could claim Denmark and Norway. His costs in Denmark were zero casualties and in Norway only about 5,300 dead or wounded. Hitler's hat size continued to grow along with hubris, arrogance, and over confidence.

CHAPTER 11

The Battle of France

One month after Hitler invaded Denmark and Norway, he executed a plan to invade and conquer France. The French and Germans had long been enemies. France was defeated by Prussia (Germany) in the Franco-Prussian war of 1870-71. France got even by defeating Germany in the Great War, (a.k.a., World War I). The onerous Treaty of Versailles which Hitler and Germany hated was a creation largely by France and the documents ending World War I were signed in France. Of all the countries Hitler invaded during the World War II era, he had a personal vendetta against the French. To accomplish his objective of defeating France, Hitler also overran Luxembourg, Holland, and Belgium. To him, they were basically collateral damage.

Hitler attacked France in 1940 for several reasons. (1) He sought vengeance against France and their arrogant imposition of the humiliating and burdensome provisions of the Treaty of Versailles. (2) Hitler's greater objective was to invade Russia and its Ukrainian region for their wheat, oil, and additional *lebensraum* (i.e., living space) for Germany. Hitler, however, understood the perils of fighting a two-front war. He wanted his western frontier with France secure from French intervention. Therefore, his invasion of France was strategic. He wanted France neutralized militarily so that he could pursue his next objective of gobbling up a large chunk of Russia and the Ukraine.

The Maginot Line

France for its part did not trust the Germans. They fully anticipated further hostilities in the future. Therefore in the twenty or so year interval between World War I and World War II, they had been busy building a line of concrete fortifications abutting the German border. It stretched from Switzerland in the south to an area of northeastern France which abutted the Ardennes Forest region. This was called the Maginot Line. It was a 280 mile succession of concrete blockhouses, pillboxes, and fortresses. The French did not continue the line through or around the Ardennes Forest because they considered it impassable for any major military operation. Moreover, they hoped that if Germany did bypass the line, they would have to invade through Belgium, thus placing the battles and destruction of war on Belgian soil rather than French.

A masterpiece of military construction, the Maginot Line was built mostly of extremely thick concrete and steel. Most of the structures were underground and connected by tunnels. In the tunnels were narrow gauge railroad tracks which could shuttle personnel, ammunition, and supplies out of sight and out of reach of the enemy. The concrete casements, block houses, and fortresses were constructed of immensely thick, steel-reinforced concrete which was impervious to artillery or aerial bombardment. Concrete or steel machine gun turrets and pill boxes were sited all along the way to provide a deadly cross fire against advancing infantry. At the border, a line of antitank blockhouses was intended to provide resistance to armored assault. The notion was to sufficiently delay the enemy to allow French forces time to fully man the line if war erupted.

The Maginot Line included underground barracks for infantry and fortress garrisons. They were heated, air conditioned, and could hold supplies for lengthy periods of time. The fortifications had their own electric generating systems and the entire line was interconnected with its own telephone system—of course, all underground. The Maginot line was thus basically impregnable to the military technology of the 1930s.

Yet, the Maginot line had two unintended consequences. (1) It lulled French leadership into a false sense of security. And (2), the Germans, well aware of the line, therefore planned to bypass it altogether when the time came.

Planning for the French invasion

Hitler wanted to invade France right after it had declared war on Germany in September of 1939. His generals talked him out of it, reminding him that his armies needed to re-furbish, refresh, and refit after the campaign in Poland just that month. Then he wanted to try in October and November. His generals were aghast and made excuses why the Wehrmacht was not ready—which it was not. In fact, several generals even contemplated a coup d' etat against the Fuhrer. Winter weather then slowed down Hitler's ardor. But by spring of 1940, he was determined to invade France. There was infighting amongst his generals over a plan of attack. But the driving force behind the planning, however, was Hitler and not his generals. They thought that war with France should be postponed for at least two years to allow the Wehrmacht time to fully mature.

By April, however, a basic plan had been agreed upon by Hitler, OKW (German military high command), and the army. The plan was relatively simple in concept. Using an American football analogy, the Germans planned to do an end-around play on France. They knew better than to try and butt their heads against the Maginot Line. They would therefore, do an end-run and by-pass it. Their basic strategy was a one-two plan.

First, they would attack through Belgium and on into France. (This of course would mean invading tiny Luxembourg which was in their way.) While they were in Belgium, they would also occupy the Netherlands to seize control of the Scheldt waterway and Europe's largest port of Amsterdam.

After defeating the Belgians, Hitler's panzer divisions would then drive south into France. Their plan was, of course, to defeat French forces. But they also wanted this initial incursion into France to be a ploy to draw the main French forces toward them and away from the Ardennes area.

Then, while the French were dealing with Wehrmacht forces invading Belgium, other German forces would attack through the Ardennes, knowing the French would be taken by surprise. Though the Ardennes Forest was not an easy area for military maneuvers, the Germans knew they could do it and they did. And so, the battles erupted.

Defeat of Belgium

On May 10, 1940, German forces invaded Luxembourg and then soon drove into Belgium. On the first day, the Luftwaffe basically destroyed the Belgian air force on the ground. One of the major defensive installations in Belgium was at the junction of the Meuse River and the Albert Canal. There the Belgians had constructed a massive reinforced concrete fortress called Eben-Emael. At the time it was thought impregnable and reputedly the largest fortress in the world. It was designed specifically to defend Belgium from German attack. However, the Germans simply landed assault gliders loaded with airborne troops on top of the massive Fortress. They quickly defeated the installation. Meanwhile, other airborne troops landed near strategic bridges and took them. This was followed by the main force of German ground troops which quickly overran Belgian forces. Though aided by both French and British expeditionary forces, the Germans simply overpowered them. One major key was that the Germans from day one had air superiority. Once again, JU-87 Stukas were insurmountable in their close air support of German ground troops. At 4 a.m. on June 2, 1940, Belgium surrendered.

On May 14, the Germans also invaded and occupied the Netherlands to the northeast of Belgium.

Of note is that Belgium and France were aware of the German build up. French reconnaissance pilots reported military convoys in Germany over 60 miles long at the German border. The Belgians took the matter seriously; the French did not. The Belgians fought heroically, but simply could not overcome German superiority in virtually every dynamic. The French army moved into Belgium to assist the Belgians against the Germans, only to be slowly pushed back into France.

Invasion of France

And so, with Belgium out of the way, the Germans rolled into northern France. The code name for this first part of their overall strategy was called *Fall Gelb* (Case Yellow). It was to lure French forces to the north and away from the Ardennes Forest area. This northern battle force was called Army Group B under the command of General von Bock.

Ardennes

Though the French thought passage of an army through the Ardennes Forest and hills to be impossible, German Army Group A, under the leadership of General von Rundstedt did so anyway, utterly surprising the French. The German advance was hampered by the number of vehicles trying to force their way along the poor roads through the Ardennes. Army Group A had 41,140 vehicles, with only four march routes through the Ardennes. French reconnaissance aircrews had reported German armored convoys by the night of May 10th but assumed them to be secondary and headed to Belgium. Nevertheless, after making their surprise push through the Ardennes, the Germans then drove into the Somme valley, cutting off the Allied units that had advanced into Belgium to meet the expected German invasion. The feint of Army Group B coming out of Belgium had worked. The French thought this was the main attack and deployed their forces accordingly. The French felt no sense of urgency regarding the Ardennes incursion as they believed the buildup of German divisions there would be correspondingly slow. Thus Army Group A along with Group B, proceeded to cut off and defeat large French forces in northern France.

France was not a paper tiger. In May of 1940, the French army fielded 110 divisions. Their heavy tank, the Char B, was superior to the light Panzer I, II or medium Panzer III tanks the Germans deployed into France. In the ensuing battles, some French units fought well, but were hampered by poor leadership, poor communications, and poor tactics.

The Germans had four distinct advantages over the French which were not only instrumental, but decisive in their victory over France.

First, they had installed two-way radios in their tanks and field commands. They thus could communicate instantaneous adjustments during battle according to the changing situation. The French communicated with their forces by written orders often delivered by runner. Their tanks and ground commanders had no radios. French tank unit commanders often had to dismount and hand-deliver orders to their unit. They also utilized a field telephone system, but in the heat of battle phone lines were often cut. German use of radios was a profound advantage over the French in allowing the Germans to adjust immediately to changing battlefield environments. The French system was slow and often the battlefield had changed by the time orders from above had filtered down through the chain of command.

Second, and more importantly, the German army also had direct radio contact with the Luftwaffe overhead. The Germans established air superiority almost immediately after entering France, destroying much of the French air force on the ground. The formidable JU-87 Stuka dive bombers were devastating to French forces. They roamed the skies above looming battles. When German forces were threatened by superior French tanks, fortifications, or infantry formations, the Luftwaffe was called in and the Stukas usually neutralized if not eliminated the French threats. This more than anything won the day for the Germans in the Battle of France. The French had no effective means of neutralizing the Stukas. Because the German ground commanders had direct radio contact with the Luftwaffe, Stukas could be and were overhead in short order to achieve battlefield advantage.

Third, early in the war, the Germans allowed and even encouraged their ground commanders to take the initiative and make decisions based on changing battlefield conditions. This gave them great flexibility to adapt to what was developing around them. The French, on the other hand, used a rigid top-down system of command, which coupled with their slow communication system left them at a distinct disadvantage. The French were essentially using World War I tactics to fight the Germans who had mastered the art of rapid advance and mobile attack. The Germans with their faster tanks, motorized infantry, and close-air support left no doubt as to the outcome of the greater battle. (Ironically, later in the war, Hitler distrusting his generals would take direct control of battlefield tactics, micro-managing the war. In the final analysis, this was a blunder and a substantial part of the coming defeat of Germany in Russia.)

Fourth, the Germans fielded relatively younger commanders who were aggressive in their tactics versus the French who were encumbered by old men in leadership who were tired, tentative, and timid in decision making. Moreover, the French army was riddled with low morale and a lack of military discipline. Again, the outcome of the battle was never in doubt.

Dunkirk

Battles between the French and Belgians against the Germans raged back and forth for two weeks with the Germans suffering some tactical defeats. By now the British Expeditionary Force (BEF) was involved in the battle. But the German ability to communicate, coordinate, and their ability to call in close-air support overcame the combined forces of the Allies. The Germans pressed westward along the Somme River valley toward the English Channel. On May 28, 1940, the Belgians withdrew from the battle.

Since declaring war against Germany in early September 1939, the British had landed approximately 390,000 troops in France. Some were engaged with the Germans as they drove into northern France out of Belgium and from the Ardennes.

As the Germans advanced, they severed all communication and connection between the northern and southern branches of Allied forces. In the process, the Germans pushed several hundred thousand Allied troops in the north into an increasingly small sliver of the French coast. As early as May 19, 1940 the British commander on the ground began contemplating evacuating all British forces back to England lest they be killed or captured by the rapidly advancing Wehrmacht. The place where the British, and some French forces, were cornered was the small French port of Dunkirk.

Hitler was aware of all of this. However, he fretted that his aggressive generals had overextended themselves, leaving their flanks open to Allied attack. Therefore, he ordered them to halt their drive to the sea—another blunder. Luftwaffe commander Hermann Goering also assured Hitler that his Luftwaffe could prevent any evacuation from Dunkirk. Hitler would rescind his order on May 26, but the British were already well underway with their plans to evacuate.

The British code name for their evacuation plan was Operation Dynamo. Virtually anything that could float in southern England was pressed into action. Water depths at Dunkirk were too shallow for British warships to dock. Hundreds of fishing boats, pleasure yachts, tugboats, lifeboats, ferries, motorized barges, and other civilian ships of every size and type raced to Dunkirk evacuating British troops back to England. Some smaller vessels ferried evacuees further offshore to deeper water and waiting British warships. Officially, at least 861vessels of all sizes and character got involved, though some estimates place the total as high as 1,200. Meanwhile, the Luftwaffe attacked the flotilla sinking 243 British vessels. However, the Royal Air Force (RAF) intervened and for days on end there were raging aerial dog fights between British fighters and the Luftwaffe. When it was over the British had lost 177 aircraft; German losses were 240 aircraft.

Operation Dynamo went on for nine days, ending on June 4. A total of 338,000 British and French troops were evacuated. Approximately 198,000 British troops were rescued and another 140,000 French soldiers as well. (Many of the French were soon re-landed further west in the Normandy region to rejoin the battle again against the Germans.) Approximately 90,000 allied troops were left behind. Additionally, *all* of the British heavy equipment from tanks to artillery to trucks was left on the beaches of Dunkirk. Approximately, 16,000 French soldiers died during the evacuation as did 1,000 British troops. On June 14, 1940, the British high command ordered the rest of its forces in France evacuated back to England. On June 25, another 192,000 Allied troops, including 144,000 British troops were evacuated from other French ports which had not yet fallen to the Germans. England had been able to evacuate about 342,000 troops of the original 390,000 they had sent to France in the first place. But the British army was exhausted and had lost the bulk of its equipment.

Left behind were huge amounts of ammunition: 880 field guns, 310 guns of large caliber, over 500 antiaircraft guns, 850 anti-tank guns, 11,000 machine guns, nearly 700 tanks, 20,000 motorcycles, and 45,000 automobiles and trucks. It would take many months for the British to resupply their army.

The Collapse of France

After Dunkirk, both the French and the Germans regrouped and prepared for the final showdown. The Germans called this phase of the battle *Fall Rot* or Case Red. Facing imminent defeat, the French army stiffened some as their commanders came to understand how to cope with German tactics. However, though battles raged for another several weeks, on June 22, 1940, France surrendered. Upon hearing of the French suing for peace, Hitler traveled to France, and selected the Forest of Compiègne as the site for the negotiations. Compiègne had been the site of the 1918 Armistice, which ended the First World War with the humiliating defeat of Germany. Hitler viewed the location as revenge. In fact, he ordered the very railroad car in which the 1918 Armistice was signed moved onto the site from a museum. There the surrender papers were signed. Hitler thus rubbed the collective noses of the French in defeat at the same place the Germans had been humiliated 22 years earlier.

Northern France, including Paris came under German occupation. Hitler allowed the southern portion of France to remain technically free, though under the thumb of the Nazis. That section was known as Vichy France. It continued until 1942 when Hitler occupied even that part of France when the Americans invaded French North Africa.

Germany lost over 27,000 soldiers killed in the invasion of France, with another 18,000 missing in action. Hitler had initially thought he might suffer as many as one million casualties in France. He was elated at his relatively low losses. The campaign had taken only six weeks, ending on June 22, 1940. (Exactly one year later on June 22, 1941, Hitler would invade Russia.) The reaction in Germany was euphoria, with a strong upsurge in warfever. Hitler's popularity soared to near unanimous heights with an official celebration of the French capitulation several weeks later on July 6. To the average German, their Fuhrer had become akin to the Messiah. Many in Germany almost worshipped him. For perhaps the first time there was a genuine war-fever among the population.

Hitler now ruled most of Europe, from the Atlantic on the west to the border with Russia on the east; from the Arctic Circle in Norway to the Mediterranean to the south. The several nations not under Nazi domination were not so secret suppliers of the Reich. Sweden profited handsomely from its sale of iron ore to Germany. Likewise, Spain sold wolfram, the basic mineral necessary to produce tungsten, needed in weapons production. The Swiss were secretly bank rolling the war effort of the Third Reich and were likewise profiting handsomely. Finally, Italy had become Hitler's ally to the south. In the Spring of 1941, Hitler would overrun the Balkans, completing his conquest of Europe.

In July of 1940, Hitler was the most powerful man in the world. He ruled Europe from one end to the other. Only pesky England remained a thorn in his side and he intended to deal with them forthwith. Once again, Hitler's hat size increased as did his hubris, arrogance, and cockiness. He had built Germany into the most powerful nation on earth. He began to further believe his own propaganda that the Aryan-Nordic race, personified in Germany, was a super race and invincible. Against the concerted advice of his generals, whom he considered not worthy of his brilliance, he had instigated and executed German success from Poland to France. He had conquered Austria, Czechoslovakia, Poland, Denmark, Norway, Luxembourg, Holland, Belgium, and now France. Others must surely recognize he was a military-political genius.

"Pride goeth before destruction, and an haughty spirit before a fall."
Proverbs 16:18

CHAPTER 12

The Battle of Britain

After the euphoria of victory over France in early July 1940, Hitler wasted no time in turning his attention to England. He had thought England would not go to war when he invaded Poland. After England declared war on Germany on September 3, 1939, Hitler tried to induce England to back out and seek a negotiated peace. No dice. He then successfully proceeded with his invasions of Denmark, Norway, Belgium, Holland, and France. But by the summer of 1940, Hitler viewed British intransigence as reason to teach them a lesson they would never forget.

The Fuhrer finally came up against a country which he could not easily roll over. He did not have the means to launch an amphibious invasion, though it certainly was contemplated. Rather, he tried to bomb England into submission. The Germans utilized twin-engine medium bombers which had barely enough range to fly from the continent to England and back. But those bombers with their fighter escorts almost brought England to its knees.

Several things saved England: (1) the tenacity, bravery, and skill of the Royal Air Force (RAF) which though stretched to almost its breaking point went head to head with the Germans and even proved superior to them. (2) Initially Hitler had focused on destroying the RAF both in the air and on the ground, bombing their air fields, their aircraft production factories, and shooting down as many by the Luftwaffe (German air force) as he could—and he almost succeeded. But he was distracted from that goal by deciding to bomb London in retaliation for English bombing of Berlin.

As early as September 1939, Hitler began to think about invading England. Those plans were shelved, however, as he proceeded to conquer Western Europe. As those plans again re-developed in mid-summer of 1940, Hitler determined to launch an amphibious attack across the English Channel against England. It was called Operation Sea Lion. However, for that to ever happen, Germany had to have air superiority over the English Channel which meant the Royal Air Force had to be eliminated. That was primarily what the Battle of Britain would become—the attempt by the Luftwaffe to destroy the RAF.

Apart from the success or lack thereof against the RAF, Operation Sea Lion in all likelihood would have failed. Germany was not a sea power. Britain was. In fact, England had the most powerful navy in the world at that time. Furthermore, Germany never developed or built the landing craft which the United States would do several years later. When the United States and England invaded France on June 6, 1944, they had thousands of craft specifically designed for amphibious assault such as the ubiquitous Higgins boats which ferried hundreds of thousands of Allied troops ashore from ships in various contested Allied landings. Nor did Germany have large specialized landing ships such as LSTs (Landing Ship Tanks) which could directly land heavy equipment—tanks, trucks, and artillery—on a contested beach.

Rather, the Germans planned to use shallow-draft river barges from the river systems of northwestern Europe to ferry troops across the channel. They would be propelled by river tug boats, making a speed of about two knots. The Germans seemed to have no nautical sense that such craft could only successfully cross the English Channel if seas were flat calm, which they seldom were. In any kind of seas, river barges would swamp and sink. Moreover, the British Navy would surely destroy the most of them in any event, even with the Luftwaffe attacking.

Three years later, the United States and Britain spent many months meticulously planning Operation Overlord (D-Day) which involved thousands of seaworthy ships. At Hitler's urging, in mid-summer of 1940, rather the Germans cobbled together their plan in short order and were utterly ill-prepared if they had actually tried to execute it. Operation Sea Lion was a disaster waiting to happen. Fortunately for Britain, it never happened. The Germans were even more fortunate that they did not attempt it, for if they had, it would have wound up as an unmitigated disaster for them.

Notwithstanding all of that, Hitler forged ahead with the air war against England. Once again, the goal was to destroy the Royal Air Force. The British recorded that battle as beginning on July 10, 1940 and lasting until the end of October of that year. The strategy of Hermann Goering, head of the Luftwaffe, was twofold: (1) to bomb RAF airfields and ground facilities into oblivion and (2) entice the RAF to come up and engage with battle-experienced Luftwaffe fighters. The assumption was that in the coming battle of attrition, Britain would lose enough pilots and planes that the Luftwaffe could then finish them off.

The British fielded two excellent fighter planes: the Hawker Hurricane and the Supermarine Spitfire. All things being equal, they could hold their own against German fighters. In fact, as the battle would unfold, the British usually got the better of it, to the chagrin and frustration of the Luftwaffe. RAF pilots certainly had their baptism of fire in defending the evacuation of forces at Dunkirk. Thus, they did not enter the Battle of Britain wholly inexperienced.

The Germans fielded two top-notch fighters: the Messerschmitt Me 109 and the Focke Wulf Fw 190. They were state-of-the-art fighter aircraft for their day. Of course, they had other combat aircraft, but the Me 109 and Fw 190 were their workhorses. Moreover, they had seen action in the Spanish Civil War, the invasion of Poland, and most recently the Battle of France. Their pilots therefore were experienced and battle hardened.

The Germans also fielded a number of battle-tested medium bombers including the Heinkel He 111, the Dornier Do 17, and the Junkers JU 88. These each were two-engine bombers capable of flying from the continent into England to drop their bomb loads. Noteworthy is that the renowned JU 87 Stuka proved almost useless as a strategic bomber against England. It was designed for tactical, close-air support of ground forces, but was relatively slow and a sitting duck for the high-performance British Hurricanes and Spitfires. Of further note, the Germans did not have any long-range, strategic, heavy bombers. As the air war would later unfold, the British and then the United States Army Air Force would unleash thousands of four-engine heavy bombers against Germany, but the Luftwaffe never produced any themselves.

Initially, the Luftwaffe began small scale raids against England even while the Battle of France was still raging. By mid-July 1940, the Luftwaffe began to attack British coastal shipping, sinking ships carrying needed war materiel. The main attack, however, against RAF installations began in Mid-August of 1940. The Germans intended to entice the RAF fighters up into the air so they could destroy them there. Meanwhile, German bombers would attack their airfields and installations. The battles became vicious. Dogfights between the RAF and Luftwaffe fighters raged across southern England and over the channel on a daily basis. The Germans were intent on not only destroying British aircraft, but eliminating England's experienced pilots. Wave after wave of attacks by hundreds of German war planes crossed the channel on a daily basis. The British pilots were for the most part young men in the late teens or early twenties.

The British, in the years leading up to the war, had developed an effective system of coastal radar as an early warning system. The Germans never fully appreciated how effective the British radar was. As waves of German bombers and fighters would rise up in the air over the continent, the British radar alerted Fighter Command which then launched fighter planes and vectored them to the incoming enemy. Without it, the British would have failed. But the toll was terrible. Though the British held their own and then some against the Luftwaffe, fighter pilots were on call day and night. Many would land and sit in their planes to sleep on the ground as ground crews re-fueled and re-armed them. Able to catch an hour or two of sleep, they were called upon again and again to go up and do battle with the Luftwaffe. Attrition became the enemy of the British. Their aircraft factories could essentially keep up with planes being lost in battle, but the RAF could not replace experienced pilots. The British lost over 1,700 planes in the battle, though the Germans lost over 2,200.

German intelligence was poor and Goering was told so many British fighters had been destroyed in battle, that in a few days, the battle would be over. Though German reports of British losses were no doubt optimistic, what Goering and his intelligence did not realize is that British industry was replacing aircraft lost in battle as fast as they were being shot down.

Nevertheless, the battle was relentless, vicious, and deadly for both sides. On August 20, 1940, British Prime Minister Winston Churchill in a speech before Parliament said, "Never in the field of human conflict was so much owed by so many to so few." He was referring to the relatively small number of RAF fighters holding the Luftwaffe at bay. The fact is that the Luftwaffe was on the verge of destroying the RAF. Its airfields were in shambles. Its pilots were exhausted. And England could not replace the experienced pilots being shot down.

During the last week of August and the first week of September, in what would be the critical phase of the battle, the Germans intensified their efforts to destroy the British Fighter Command. Airfields, particularly those in the southeast, were significantly damaged but most remained operational. On August 31st, Fighter Command suffered its worst day of the entire battle.

Then, something happened which changed the course of the battle and ultimately the war. It could have been a stroke of good fortune for the British, or perhaps Divine Providence intervened. Hitler had directed that the Luftwaffe bomb all things pertaining to the RAF. They were explicitly ordered to avoid bombing London. Yet on the night of August 24, 1940, German Heinkel 111 bombers, perhaps unintentionally, bombed central London. The city was blacked out and they were not sure where they were, so they just dropped their bombs. The next night, German bombs also fell on the outskirts of London.

In retaliation, Churchill ordered British heavy bombers to attack Berlin. Hermann Goering had guaranteed Hitler and the German people that British bombs would never fall on Berlin. On August

25th, the RAF launched a bombing raid on Berlin in retaliation. The Germans were shocked and Hitler was infuriated.

On September 7, Hitler ordered 300 bombers to attack London in the first of 57 consecutive nights of bombing. Losses to civilians were devastating. Hitler hoped his terror campaign against London would so demoralize the city, that they would go mad and force the British government to sue for peace. In the ensuing months, as many as 43,000 British civilians were killed by the German blitz. However, to put that in perspective, London in 1940 had a population of 8.6 million. The percentage killed was about one half of one percent of the population. Nevertheless, many in London sent their children out of the city to be taken care of by government sponsored child care in the countryside. Moreover, two million homes were damaged or destroyed.

The damage was essentially of terror. The Germans accomplished little strategically or of military value, but they did indeed terrorize the population of London. Yet, in response, Londoners shook their collective fists at the German raiders, dug in, and went about their lives. The Blitz would go on in some degree until early May of 1941, though for all practical purposes it ended in October 1940.

However, what seemed to be a disaster was actually a blessing in disguise. When Hitler shifted his focus to London, he at the same time moved away from destroying the RAF. The bombing of London would in fact be a strategic blessing for England and another huge blunder by Hitler. The RAF had time to repair and rebuild its airfields. Its front-line fighter pilots, though involved in defending London were now engaged in shooting down bombers rather than dog fights to the death with German fighters. Meanwhile, pilots from the British Commonwealth began arriving in London to join the battle. They came from New

Zealand, Australia, Canada, South Africa, Rhodesia, as well as from Belgium, France, Poland and Czechoslovakia. There were even some pilots from the neutral United States and Ireland. The British pool of experienced pilots was stabilized and actually increased.

On September 15 Fighter Command of the RAF repelled a massive assault by the Luftwaffe, inflicting severe losses that were becoming increasingly unsustainable for the Germans. Although German raids would continue for several more weeks, it had become clear that the Luftwaffe could not achieve the air superiority needed to invade England. Hitler indefinitely postponed Operation Sea Lion. His naval commanders persuaded him that launching an amphibious attack against England any later in the year would fail as autumn storms were the norm on the English Channel.

Though Britain had suffered terrible blows during the Battle of Britain and particularly the Blitz of London, they had thwarted the German strategic plan of invasion. For the first time in World War II, Germany had suffered a decisive defeat. The Luftwaffe had suffered massive losses with over 2,200 first-line aircraft lost in the battle. The British on the other hand had managed to replenish the 1,700 aircraft it had lost in battle. In fact, because of British production, the RAF actually had more fighters at the end of the battle than when it began. Germany could not crush the RAF. The Luftwaffe, with its lack of heavy bombers and failure to fully identify critically important targets, never inflicted significant *strategic* damage on England. Moreover, internal weakness in the industrial infrastructure of Germany was exposed. The Luftwaffe suffered from constant supply problems from the inability of Germany industry to produce the aircraft needed to win the battle. This pointed to failed leadership at the top. Hitler had given Hermann Goering a free hand with the Luftwaffe. His failures were apparent.

The failure of Germany in the Battle of Britain was a turning point in the war. England had shown that a nation determined to fight Hitler could withstand his onslaught. He was not invincible and the German military were not supermen or a superior race. Though Germany would continue to send bombing raids against England for the next several years, they were more of a nuisance than having any strategic importance. Hitler had been stopped. Victory in the Battle of Britain did not win the war, but it made total victory possible later. Less than four years thereafter, the Allies would launch the invasion of Nazi-occupied France on D-Day from English soil. That in turn would prove decisive in ultimately bringing the war against Germany to an end.

Unfazed, Hitler began to make other plans.

CHAPTER 13

The Evil Empire

"And the great dragon was cast out, that old serpent, called the Devil, and Satan, which deceiveth the whole world: he was cast out into the earth, and his angels were cast out with him," Revelation 12:9.

Across the centuries, there have been great conquerors such as Alexander the Great, Julius Caesar, Genghis Kahn, and more recently Napoleon Bonaparte. Some were cruel; some were not. Adolf Hitler indeed was a conqueror, but he was more than cruel, he was evil incarnate. It is the contention of this author, that Hitler was empowered and used by the Devil himself. As a young man, Hitler had gotten into the occult. In turn, the occult likely had gotten into him. The occult is a portal into the spirit world and the realm of the Devil in particular. In making contact with the occult, Hitler was at the least influenced if not possessed and empowered by a satanic spirit, possibly the Devil himself.

"Then entered Satan into Judas surnamed Iscariot,"
Luke 22:3.

Secular writers avoid anything to do with the spirit world, possibly because of how they were educated or because they have no understanding of it. There have been numerous books and papers written trying to explain how and why Hitler turned out as he did. He is psychoanalyzed—how his father treated him, that he was a loner, he had unusual charisma, or speculation about his sexual proclivities. But the world is full of young men who did not have a good relationship with their father or were loners in life. What was unique about Hitler, from his meteoric rise from obscurity to great power as well as his absolute obsession against the Jews, indicates an evil and supernatural power within him.

> "*But unto you I say . . . which have not known the depths of Satan,*" Revelation 2:24.

Hitler's infamous volume, *Mein Kampf*, sets forth his malevolent obsessions. Russia would have done well to take it seriously, for he spelled out in considerable detail his plans to conqueror western Russia and the Ukraine. But Hitler's malevolence was much deeper than being a cruel conqueror, which he was. His entire world view was a deranged obsession with his acrid ideas of racial issues.

Hitler believed that Germany and the Nordic countries were descended from a mythical Aryan nation, superior to all other racial groups. Among other things, Hitler's prototypical Aryans had blond hair, blue eyes, and were tall and handsome. (Ironically, Hitler had none of these characteristics.) The Aryan Germans allegedly were of superior intellect, superior strength, and superior culture. He was intent on maintaining the purity of that alleged race in the German people.

On the other end of his spectrum were the Slavic nations to the east, including Russia, Poland, Ukraine, some of the Balkan states, and Gypsies from Romania in particular (called *Roma*). They were referred to as *untermencsh*—lower humans, or subhuman. To Hitler, the latter were meant to be enslaved and corralled, and not allowed to cohabit with Aryans, *ubermensch*—superior humans—thus diluting the purity of Aryan stock.

But above all else, Hitler had a pathological hatred for the Jews. To him, they were racial vermin worthy only to be exterminated. Anti-Semitism was not unique to Hitler. It was a relatively widespread prejudice across Europe particularly in Germany in the late 19th and 20th centuries. For example, Hitler blamed the onerous Treaty of Versailles of 1918 on Jewish influence upon the French and British leadership. He hated the Bolsheviks, not only for their communist ideology, but because some of the leading Bolsheviks were indeed Jews. In fact, Hitler blamed just about anything bad in the world on the Jews—from child pornography to crooked businessmen to corrupt government leaders.

Hitler's obsession against the Jews went beyond alleged or perceived political and financial corruption. To his warped values, the Jews were not only *untermensch*, but despicable, and destined only to being annihilated.

> *"And when the dragon saw that he was cast unto the earth, he persecuted"* them. Revelation 12:13

There is another being who has a similar hatred of the Jewish people—the Devil. Almost four thousand years ago, God promised to bless and prosper the descendants of Abraham (Genesis 12:1-3). His grandson was called Israel. The Jewish people to this day are descended from Israel through Isaac, and thence Abraham. Moreover, God has promised that one day, Israel will be the capital nation of the world when the Jewish Messiah returns to rule and reign. The Devil, of course, is aware of this and over the centuries has brought great persecution, pogroms, discrimination, prejudice, and misery to the Jewish people. The latter term is another name for the descendants of Judah, who was one of the sons of Israel. There is a distinction between the terms *Jew* and *Israel*. But for all practical purposes, they are synonymous, at least in the context of modern history. In short, Satan hates the Jewish people and the nation of Israel. Because they are God's chosen people, Satan has endlessly attacked and persecuted them in defiance of the Almighty.

> *"And then shall that Wicked be revealed . . . even him, whose coming is after the working of Satan with all power,"* II Thessalonians 2:9.

In Adolf Hitler, the Devil found a man eager to exterminate the Jewish race and thus be an affront to God.

Another focus of Hitler, detailed in *Mein Kampf,* was expanding the territories of Germany, particularly toward the east to allow greater *lebensraum*—living space. Hitler intended to conqueror western Russian and the Ukraine region for precisely that reason. He further intended to either eliminate the *untermencsh*-Slavic nations or at the least enslave them. He thereupon would relocate millions of his Aryan-German people onto this expanse of fertile and productive land.

Coincidentally, the bulk of Europe's Jews just happened to live in the regions to the east of Germany. By invading and conquering Poland, Russia, and Ukraine, Hitler would be killing two birds with one stone. He would be appropriating the lands he coveted and at the same time rounding up the sizeable Jewish population living there and exterminating them.

*"Let no man deceive you by any means: for that day
shall not come, except . . . that man of sin be revealed,
the son of perdition,"*
II Thessalonians 2:3.

Hitler was *not* the prophesied Anti-Christ, but very well may
have been a *prototype* thereof.

Upon coming to power in Germany in 1933, Hitler wasted no
time in beginning to deal with what he called the "Jewish question." It started relatively slowly and reached its peak in the
1940s as Germany in fact invaded the lands to the east. Though
Jews represented less than 1% of the German population in 1933,
Hitler immediately began dismissing non-Aryans from civil service, liquidating Jewish-owned businesses, and stripping Jewish
professionals of their clients. In 1935, a series of laws were
passed called the Nuremberg Laws. Any individual with three
Jewish grandparents was considered a Jew, while those with two
Jewish grandparents were regarded as *Mischlinge* (halfbreeds).

Under the Nuremberg laws, Jews were targeted for persecution and stigmatized. In November of 1938, this hatred erupted
in what has come to be known as *Kristallnacht*, or the "night of
broken glass." Jewish synagogues in Germany were burned and
windows in Jewish shops were smashed all across Germany.
Over 100 Jews were killed that night and thousands more were
arrested. From 1933 to 1939, hundreds of thousands of Jews left
Germany. Those who remained lived in a continual state of un-
certainty and terror.

As World War II began with the German invasion of Poland, the
Nazis followed right behind the Wehrmacht. The military arm of
the Nazis was called the SS (*Schutzstaffel*, or Security Squads). It
was originally established as Adolf Hitler's personal bodyguard

unit, but as the war developed, the Waffen-SS (i.e., the armed SS) became a military force parallel to the Wehrmacht and answerable directly to Hitler. SS military units were the elite of German forces. They enlisted only Nazis sworn to loyalty to Hitler and who gave the appearance of being full-blooded Aryans. Furthermore, SS military units were given the best equipment and priority over ordinary Wehrmacht units. Not only did the SS fight alongside the Wehrmacht, they became the agents of implementing Hitler's racial policies.

There were two main groups within the SS: the Allgemeine SS (General SS) and the Waffen SS (Armed SS). The Allgemeine SS was responsible for enforcing the racial policy of Nazi Germany and general policing; whereas the Waffen SS were combat units parallel with the Wehrmacht.

Within the Allgemeine SS was the *Sicherheitsdienst* otherwise known as the SD. The term meant "security service." It was the Nazi secret intelligence service, seeking out any enemies of the state. They were one of the agencies which actively hunted down Jews, particularly as the Wehrmacht marched eastward into Poland and later Russia. The SD became the shock troops of enacting the holocaust, particularly on foreign soil while the Gestapo did so within Germany proper. But within the SD was an organization called the *Einsatzgruppen*. They were paramilitary death squads that eventually became responsible for the mass killings of the Jews. The head of the SD was Reinhard Heydrich. At the end of July 1941, he was given written authorization to ensure a governmentwide effort to implement the so-called "Final Solution" in all territories under German control. The "Final Solution" was the Nazi code word for the elimination of the Jews.

The SD followed close behind the front-line operational combat units of the Wehrmacht and Waffen SS as they invaded one country after another. This began as early as the occupation of Austria and Czechoslovakia. They immediately began rounding up and hunting down Jews. Initially in Poland, the Jews were lined up before shallow graves and mowed down with machine guns. As that proved to be an inefficient means of killing, poison gas would soon be instituted.

"The power of Satan," Acts 26:18

In 1941, Jews from all over Europe, as well as hundreds of thousands of European gypsies, were transported to Polish ghettoes which were essentially urban prisons. As the death camps were brought on line, the Jewish ghettos were emptied and the inmates thereof transported to various death camps. When Germany invaded Russia in the summer of 1941, the SD, following the frontline troops, murdered over 500,000 Jews.

And so on July 31, 1941, Hermann Goering issued a memorandum to Reinhard Heydrich, head of the SD for an *Endlösung* (final solution) to "the Jewish question." Meanwhile, the Nazis were developing a vast network of concentration camps. Shortly after Hitler and the Nazis came to power, concentrations camps began to be built in Germany, initially for political prisoners, but soon for undesirables such as Jews. The first was at Dachau in southern Germany in 1933. From 1933 to 1945, Nazi Germany operated numerous concentration camps in Germany and in other parts of German-occupied Europe. However, to implement the "Final Solution," the Nazis established six primary extermination camps on Polish soil built specially for that purpose. These were:

Chelmno (in operation from December 1941 to January 1945)
Belzec (in operation from March through December 1942)
Sobibór (in operation from May 1942 to October 1943)
Treblinka (in operation from July 1942 to August 1943)

Majdanek (in operation from September 1942 to July 1944)
Auschwitz-Birkenau (in operation from March 1942 to January 1945).

The largest complex was Auschwitz-Birkenau.

From, 1941 until the end of the war, at least six million Jews were "exterminated" in the camps. Jews were rounded up from all over German-occupied Europe and sent to the camps, usually sealed in boxcars in trains. The conditions were horrific. Many died in transit. Those who survived were separated into groups: those useful as slave labor and the rest to be eliminated.

The Nazis devised more efficient means of killing than simply lining people up and mowing them down with machine guns. Rather, they built massive death chambers resembling mass shower facilities. They eventually utilized a poison gas based on the chemical Zyklon B, the trade name of a cyanide-based pesticide invented in Germany in the early 1920s. When train loads of Jews arrived at Auschwitz, those still alive were usually in fetid condition after their ordeal of transit. As the survivors tumbled out of the boxcars, they were therefore directed to the "shower" facilities, ordered to disrobe—men, women, and children—and locked within. The poison gas then showered down upon them and in moments, they were all dead. Slave laborers then removed their bodies to mass crematoriums where their remains were incinerated. At Auschwitz alone, more than two-million people were murdered in this process which resembled a large-scale industrial operation. In smaller camps, the bodies were tossed like logs into shallow mass graves by slave labor.

Ever since, the attempt to annihilate the Jewish people has been known as the Holocaust. The English word *holocaust* comes from two Greek words *holo* and *kostos* and together they mean "totally burnt." Jews tend to use the Hebrew word *shoah* which means "utter destruction" or "catastrophe." Hitler murdered not only six million Jews, but an estimated five million others for racial, political, and ideological reasons. Especially sad, more than one million of his Jewish victims were children.

The holocaust built in intensity throughout the war, only waning as allied forces began overrunning the concentration camps. The camps ended in May 1945 when the Allies crushed Nazi Germany. To the end, Hitler blamed the horrors of World War II on "international Jewry and their helpers—the universal poisoners of all people."

Though Hitler had many fervent anti-Semites in the Nazi Party bureaucracy, from the Gestapo to the SD and the SS in general, Hitler was the motivating force behind it all. Before World War II, there were an estimated 15 million Jews in the world, with approximately 9.5 million living in Eastern Europe—Poland and Russia having the most with well over five million combined. Hitler killed about six million European Jews during the course of the Holocaust. He was not only a cruel conqueror, he was evil incarnate—a man empowered by the Devil.

CHAPTER 14

A Catastrophic Blunder

After being thwarted in his plan to invade England or at least neutralize its air force, Hitler turned his eyes to the east. Though he had surely made errors of judgment and some tactical blunders in the several years he was overrunning Western Europe, Hitler was now about to make a catastrophic blunder which along with his second major blunder would ensure his defeat.

To this point, Hitler had rolled over Europe with minimal losses and insignificant opposition. He had had virtually nothing but success. He had largely masterminded it all. His generals for most of these campaigns were hesitant, reluctant, and certainly were not the instigators of the actions. Hitler was. He came to view himself as a military genius. His hat size increased significantly. (Pride goeth before destruction.)

In every campaign from the occupation of the Rhineland to the defeat of France, Hitler's armies were victorious with minimal losses. Hitler began to believe his own propaganda about the German people being a master race and superior to other nationalities. He came to view his armies as invincible. From that developed arrogance, hubris, and over confidence.

The fact is, the Wehrmacht was indeed superior to any ground army in the world at the beginning of World War II. They had superior leadership, superior discipline, superior organization, superior morale, superior weapons, and superior tactics. On a unit by unit comparison, few infantry units in World War II were superior to a comparable German army unit. But other factors Hitler had not contemplated led to his defeat including the naval superiority of England and the United States as well as the eventual air supremacy of Britain and America over Europe. Together, they more than overcame German superiority in ground forces. In the case of Russia, the Red Army would eventually overcome the Wehrmacht by sheer brute force—overwhelming them with almost endless numbers of infantrymen.

By the fall of 1940, Hitler occupied and controlled most of Europe from France to the border with Russia and from Scandinavia to the Mediterranean. For good measure, he also invaded Yugoslavia and Greece in April of 1941.

> *If Hitler had stopped there, Europe today very well may still have been part of a German Reich (i.e., empire).*

To this point, Hitler had had nothing but success. He had conquered Europe and was its sole ruler (fuhrer). He was idolized by most of his own nation. He had fame and some wealth. His book *Mein Kampf* had made him a millionaire when that term meant much more than it does today. He was the most powerful man in Europe, in fact in the world. Only a handful of countries in Europe were not under his oppressive rule.

Neutral Switzerland was Nazi Germany's secret banker. They financed his war. Moreover, when American bombers targeted the German ball-bearing factories around Schweinfurt, the Swiss happily helped re-supply German supplies, with a handsome profit of course

Spain was a friendly neutral whom Hitler tried to persuade to no avail to be an ally. But in any event, the Spaniards supplied Germany with the relatively rare mineral wolfram from which tungsten steel was made.

Italy was his ally. Romania and Hungary would become reluctant allies, mainly out of fear of Russia.

Yugoslavia and Greece were overrun by Hitler in April of 1941. The greater point is by the Spring of 1941, Hitler was the unquestioned fuhrer—ruler—of Europe. He had accomplished that in a little over two years.

Background to the Russian Invasion

In his book, *Mein Kampf*, as early as 1925, Hitler had outlined his plan to invade Russia and neighboring regions—in particular the Ukraine region of the Soviet Union. His purpose was to secure *lebensraum* (i.e., living space) for the benefit of Germany for generations to come. Western Russia and the Ukraine especially were the bread basket of Eastern Europe. Hitler further foretold in *Mein Kampf* that Germany's destiny was "to the East." Moreover, Hitler hated the Bolshevik revolution and the ensuing Soviet Union. Conquering Russia would not only provide his coveted *lebensraum* with its rich farm lands, but would also eliminate the Communist Party from the face of the earth. Also, as noted earlier, millions of Jews lived in western Russia and the Ukraine. By conquering these lands, he could eliminate them as well.

Russia and Germany had long been uneasy neighbors. They had been enemies in World War I. After the fall of the Tsar and the Romanov Empire to the Bolsheviks in 1921, the Union of

Soviet Socialist Republics (USSR) emerged. It was the first fully Marxist-Communist state in history. The first leader and dictator of the USSR was Vladimir Lenin. Upon his death in 1924, Joseph Stalin forcibly seized power. He was a brutal dictator seeking to force collectivization upon the largely agrarian Russian population. Through bungled socialist policies of forcibly collectivizing Russian farms into state-run collectives, millions of Russian peasants died. Millions more died in the famine of 1932-33.

Stalin further sought to eliminate all political enemies or even potential ones by a series of murderous purges between 1936 and 1938. He was paranoid. Part of his purges was the officer corps of the Red Army. By 1939, Stalin had eliminated most upper-level officers from the Soviet armed forces. His fear, of course, was that they had the power to overturn him and he therefore simply eliminated them. Many were executed outright, others were sent to prison camps in Siberia called Gulags. All of this would have profound implications and consequences when Germany invaded Russia. When Russia found itself at war with Germany in 1941, Stalin was forced to recall most of the surviving officer corps and re-instate them in his headless army. Though the USSR was immense in area, it was only twenty years old in 1941 and was not a mature or stable nation.

German-Russian Nonaggression Pact

On August 23, 1939, about one week before Hitler invaded Poland, Nazi Germany and Soviet Russia announced the signing of a German-Soviet Non-Aggression Pact. To a stunned world, they agreed to take no military action against each other for the next 10 years. The treaty was negotiated by Hitler's foreign minister Von Ribbentrop and the Soviet Foreign Minister Molotov in the Kremlin. Stalin knew the Red Army was not ready for war against Germany. Hitler, for his part wanted his eastern frontier secure while he engaged in his conquests to the west, after devouring Poland. Hitler had no intention of keeping his word—he was a liar. Once France and hopefully Britain were out of the picture, he then intended to invade the Soviet Union. But the non-aggression treaty would serve his purpose in the short term.

Though the treaty was publicly announced to the world on August 23, 1939, there was a secret protocol included in the treaty which was not known until after the war was over in 1945. In that secret section was an agreement between Hitler and Stalin that when Hitler invaded Poland, the Soviets would be allowed to seize eastern Poland as well. According to the agreement, Romania, Poland, Lithuania, Latvia, Estonia, and Finland were divided between German and Soviet "spheres of influence." To the north, Finland, Estonia, and Latvia were to be part of the Soviet sphere. Poland was to be partitioned with the areas to the east going to the Soviet Union—Germany would occupy the west. Romania would go to the Germans and become a very uneasy ally of Germany in the upcoming war.

Thus, both cruel dictators, Hitler and Stalin, were satisfied. Hitler ensured his eastern frontier was secure from the Soviets while he rampaged across Western Europe. Stalin was happy to gobble up the small Baltic States and add them to the Soviet orbit. Though Stalin did not agree with Nazi policy or philosophy, he admired a strong dictator like Hitler. That personal entente would only last until June of 1941.

In light of what was to take place in the latter half of 1941, there was another amazing agreement in their treaty. Stalin agreed to sell Germany immense quantities of needed strategic materials such as coal, gasoline, wheat, along with assorted other commodities. The USSR in that era was strapped for cash and was happy to sell its commodities to Germany. Hitler was happy to purchase what it needed. This went on for almost two years with train load after train load of commodities needed to prepare Germany for war. In fact, when Hitler did actually attack Russia, a train load of Russian commodities had crossed the border just two hours before Hitler struck. Lenin had once quipped, "The capitalists will sell us the rope with which we will hang them." It was an irony likely lost on Stalin that he was selling Hitler the rope with which Germany would soon hang the Russians.

Through the Autumn of 1940 into the Spring of 1941, Hitler and his war staff planned in detail the upcoming invasion of Russia. German armed forces likewise spent this season in preparing men and materiel for the upcoming battles. In December of 1940, Hitler rejected his staff's plans and came up with his own. Hitler's plan was to be called "Operation Barbarossa."

CHAPTER 15

Operation Barbarossa

Operation Barbarossa was named by Hitler after the Holy Roman emperor Frederick Barbarossa who ruled Germany from A.D. 1152 to 1190, seeking to establish German dominance of Europe. It was the code name for the German invasion of the Soviet Union, which started on Sunday, June 22, 1941. American students of World War II in Europe usually think in terms of D-Day, the Battle of the Bulge, and perhaps the American landings in Africa, Sicily, and Italy. Operation Barbarossa, however, would dwarf all of those battles. It in fact would be the largest single military operation in world history, to that time.

On June 22, 1941, three million German soldiers in 150 divisions, with 3,000 tanks, spread over almost 1,600 miles from the northern Baltic region to the Black Sea invaded Russia. The Soviet Union in 1941 was roughly 60 times larger than Germany in land mass. It had more than twice the population of Germany (180 million to about 80 million). Those numbers would come to haunt Hitler.

But Hitler *underestimated* the Russians. He underestimated the resolve and determination of the Russian people to rise up and defend their homeland unlike the French which essentially rolled over and quit after six weeks.

He *underestimated* the Russian industrial-military complex, which once fully mobilized far out-produced war materiel than did Germany. Though Germany generally had superior weaponry compared to Russia, Hitler did not anticipate the production of the T-34 Russian tank which, all things being equal, was the premier tank in World War II.

Hitler *underestimated* the depth of the Russian infantry reserves. Though the Germans killed or captured between four and five million Russian soldiers in the first five months of the war, the Russians simply mobilized millions more. Though the Germans slaughtered millions of Russian troops, Russian numerical superiority eventually overran the Germans.

Hitler *underestimated* the immensity of Russia—it covers seven times zones. It was one thing to supply the Wehrmacht in Poland or Holland, where distances were measured in scores of miles. It was another thing to keep the German armies supplied in Russia where distances were measured in hundreds into thousands of miles. Moreover, much of the Russian rail network was of a different gauge than that of Western Europe. Hitler's generals may have known that, but such trivial details seemed to have escaped Hitler. To utilize captured Russian railroad tracks, the Germans had to unload German trains and reload onto Russian box cars or laboriously re-lay the tracks. Railroad-track repair units were at the bottom of Hitler's list of priorities.

Hitler *underestimated* the Russian weather. The Russians would quip their two great allies were General Winter and General Mud. Germans units would drive to within sight of Moscow when the Russian winter descended upon them. No one from

Hitler on down had made preparation for that and the German army almost froze to death that first winter. In the Spring when the countryside and dirt roads thawed out, they became a sea of glutinous mud through which German motorized vehicles could not move. The Russians lived there and knew how to cope with their winters and mud. The Russian winter did not defeat the Germans, but it certainly stopped them. Over the next six months of the winter and spring of 1942, the Russians had time to catch their breath, regroup, begin to fully mobilize, and ramp up armaments production. The Germans never made much substantial progress into Russia thereafter.

Hitler simply *underestimated* Russia. Though the Wehrmacht sliced through Russian troops like a hot knife through butter in 1941, the Russians were able to regroup. Hitler had told his generals that Russia was so rotten that all the Germans had to do was kick down the door and the whole structure would collapse. That seemed true for the first several months of the invasion, but certainly changed as time wore on. Ironically, the conventional wisdom of the western powers' intelligence services was that should a war break out between Germany and the Soviet Union, the Red Army would quickly collapse.

Over the next two-and-one-half years, Russia was able (with help from the United States) to mobilize millions of reserves, train new and effective officers, and fully deploy them all. By the summer of 1944, the Red Army began to slowly, but inexorably push the Germans back into Germany and ultimately on to Berlin in the spring of 1945.

Hitler's decision to invade Russia was a catastrophic blunder from which he never recovered. He was motivated in part by ignorance of how formidable Russia would be. But he was even more motived by his own arrogance. He was blinded by his own hubris and misled by his own ideology. He came to believe his own propaganda—the superiority of his Aryan army. Not only did he make a catastrophic strategic blunder in invading Russia in the first place, he added to his woes by making numerous tactical blunders along the way which he forced upon his generals. That in itself probably lost the battle for Russia. He was not a military genius.

June 22, 1941

Hitler had been amassing troops and heavy armor all along the Russian frontier for weeks prior to his jump-off day. Various foreign leaders had warned Stalin that Hitler was about to attack. By the third week of February 1941, 680,000 German soldiers had been assembled in areas on the Romanian-Soviet border. By the day of attack, June 22, 1941, over three million German troops had been assembled along a 1,600 mile front stretching from Norway to the Black Sea. When the order came, they and three thousand Panzer tanks crossed the border into Russia.

Actually German strategists had planned to invade Russia in early May, 1941. However, Benito Mussolini, the fascist dictator of Italy, had invaded Greece in October of 1940. The Greeks rose up to defend their country and things did not go well for the Italians. Hitler, fearing British intervention to help the Greeks (which they did), therefore launched an attack through Yugoslavia and down into Greece to bail out Mussolini in April of that year. By the time the Greek campaign wrapped up, the invasion date of Russia had to be pushed back about six weeks to June 22, the first day of summer. However, the delay would prove to be disastrous for the Germans later that year. The Russian winter came a little early in 1941 and stopped the Germans short of Moscow. If they had jumped off as planned in early May, things might have turned out much differently.

At 3:15 a.m. on June 22, 1941, Hitler commenced the invasion of the Soviet Union with the bombing of major cities in Soviet-occupied Poland and an artillery barrage on Red Army defenses on the entire front. Hitler and his staff had organized the German invasion into three army groups: Army Group North drove northeastward into the Soviet Union with its ultimate goal of sacking Leningrad (formerly known as St. Petersburg). Army Group South launched an attack southeastward with the goal of conquering the Ukraine region of the Soviet Union. Army Group Center attacked directly eastward with the possible objective of taking Moscow, though Hitler would equivocate on that goal for several months, to his peril. The German army, along with two Romanian divisions, smashed across the border into Russia with a total of 152 divisions.

Stalin and Stavka (i.e., the Russian military high command in Moscow) had initially ignored numerous warnings of German treachery and their impending invasion. They simply did not want to believe what was about to happen. Stalin had hoped his non-aggression pact with Germany would prevent Hitler from attacking. Stalin knew the Red Army was not ready to face the Wehrmacht. Finally, late the night of June 21 into the wee hours of the 22nd, the Soviet military districts in the border area were alerted to be combat ready, but under no circumstances to engage in provocative action against Germany. However, before those orders could filter down to the field commanders, Germany had already attacked.

As Hitler and his three-million-man army invaded Russia, they came not as liberators, but as conquerors. The Russian peasantry hated Stalin. Millions had died directly through his purges, or indirectly through his bungled attempt to collectivize Russian agriculture. The resulting famines had literally starved millions of peasants to death. They hated Stalin. If Hitler had marched into Russia with a sympathetic attitude toward Russian citizens, they would have joined allegiance to him to rise up and help him overthrow the hated Communists and Stalin.

However, Hitler was driven by his ideological obsession of subjugating the Slavic peoples of Western Russia and exterminating as many Jews along the way as possible. The peasants quickly came to the conclusion that he was worse than Stalin and threw their lot in with Comrade Stalin to defend their country against the atrocities the Nazis were visiting upon their land. The Germans burned the Russian villages, robbed and looted their granaries and livestock, shot anyone who got in the way, and simply enslaved or terrorized the civilian population as they drove eastward. The Russian peasantry rather than rising up against Stalin, became bitter and tenacious resistance fighters in the German rear. They disrupted the tenuous German supply lines and harassed German installations at night. They became a significant distraction and drain on German military resources. Hitler blundered not only by invading Russia in the first place, but also in how he ordered His troops to deal with the Russian civilian population. The latter in itself could have ensured a different outcome of the war.

However militarily, everything went like clockwork for the Germans in the first weeks of the invasion. German armies drove deep into Soviet territory. Their panzer armies enveloped large Russian forces at Minsk and Smolensk. German armored spearheads managed to drive two-thirds of the distance to Moscow and Leningrad. At the time of Hitler's invasion of Russia, German combat effectiveness had reached its zenith in the quality of its officers, training, tactics, weapons, and fighting ability. The Wehrmacht invading Russia was likely the finest army of the twentieth century.

The Luftwaffe attacked Soviet troop concentrations, supply dumps, and airfields. Additional air attacks were carried out against Russian command and control centers to disrupt the mobilization and organization of Soviet forces. The Luftwaffe destroyed over 3,900 Russian aircraft either on the ground or in the air during the first three days with a German loss ratio of about one percent. By the end of the first week, the Luftwaffe had achieved air supremacy over the battlefields of all the army groups. Slowly, however, the Russians began to get their equilibrium. By July 5, they had destroyed 491 Luftwaffe aircraft with 316 more damaged. The Luftwaffe was thus left with about 70 percent of the strength it had at the start of the invasion.

Nevertheless, the Russians had been caught flatfooted and initially were utterly unprepared for the onslaught of the Wehrmacht and the Waffen SS. Stalin's purges of the officer corps had left the Red Army in disarray. German panzers cut through Russian defenses easily. Germany had rolled over France in six weeks. Hitler forecast it therefore would take only eight weeks to bring about a collapse of the Red Army. In the first month of the campaign, it looked like he would be right. Not only was there over-optimism, but again more of Hitler's hubris and arrogance. Such, however, would bring serious strategic dislocation later in the year. No plans had been made for winter operations.

As the German Army Group North attacked toward Leningrad, the Soviets launched a substantial counterattack against them. Yet, the Germans routed the poorly led, poorly armed Red Army. Similar battles took place to the south in the region of Ukraine as well as in Belorussia. The Soviets counterattacked the German spearhead, only to be slaughtered by the superior German forces.

Hitler had more interest in taking Leningrad and later Stalingrad than he initially did of Moscow. Perhaps, it was the symbolism of the iconic names of those two Russian cities. German forces would eventually reach the outskirts of Leningrad only to be denied total victory. They in turn besieged the city from September of 1941 until January of 1944. Over one million Russian civilians died from starvation, hypothermia, and disease.

Meanwhile, German forces surrounded massive Soviet forces near Kiev to the southeast on September 16. The Germans sacked four Russian armies as a results of the encirclement battles of Kiev in September and then at Bryansk-Vyazma in October, each netting over 600,000 prisoners respectively. Sadly, most Russian prisoners of war died in German POW camps. The Germans for the most part simply let them starve to death. This again would come to haunt the Germans when the Russians gained the upper hand several years later.

The Battle of Moscow

Hitler dallied for several months about whether to order his Army Group Center to drive on and take Moscow or divert some of its forces to the northern or southern commands. Finally, on September 30, Hitler ordered Army Group Center to march toward Moscow in a campaign called Operation Typhoon. After Kiev, the Red Army no longer outnumbered the Germans and there were no more trained reserves immediately available. But the Russians had several powerful generals who would stop the Germans in their tracks. One was General Georgi Zhukov, a tough, battle-hardened, old-school Russian commander. He had led Soviet forces against the Japanese in their recent ongoing border wars with the USSR. He in fact may have been the most effective overall commander of World War II. He was one of few Russian officers who dared to contradict Joseph Stalin to his face on strategic matters. Yet, Stalin came to trust him and would receive his advice. Zhukov established a defensive system between Moscow and the oncoming Germans.

Meanwhile, Stalin had been assured by a Russian spy living in Japan that the Japanese would turn south to fight the Americans and British instead of north against Siberia. Stalin therefore transported across the Trans-Siberian railroad 18 divisions of battle-hardened troops who had been fighting the Japanese in the Far East. They were trained, experienced, and well equipped for operating in harsh winter conditions of which they were accustomed. This would prove to be a decisive advantage to the Russians.

The Soviets fielded more than a million soldiers and a thousand tanks at Moscow. They dug into multiple defensive lines. The Germans in contrast still had two million soldiers, though not all were directly involved in the attack on Moscow. They intended to conduct a series of pincer movements to surround and destroy the Red Army in front of Moscow, and then roll on into the capital. The fast moving panzers would provide the pincers and then as the infantry moved up, they would destroy the surrounded Red Army. In the initial battles, more than 500,000 Russian troops were killed or captured. It looked grim for the Russians.

But then, Russia's other two great generals intervened to slow and then stop the German armies. Few roads of western Russian were paved, the vast majority were simply dirt. Heavy rain and melting snow fell in early October, bringing with them the infamous *rasputitsa*. The muddy season turned the Russian roads and countryside into a sea of glutinous mud, causing German vehicles to sink to their axles. General Mud had arrived. He also would return in the following spring. German combat forces were stopped in their tracks. Their armored vehicles as well as their supply trucks were simply stuck on a massive scale. The

Russians knew this was coming and knew how to cope with the mud. Zhukov therefore ordered one counter attack after another, leaving the Germans battered and exhausted.

With the coming of November came colder weather which caused the mud to freeze. This allowed the Germans to again continue their advance on Moscow. By the end of November, German reconnaissance units were just 12 miles from Moscow, so close they could see the church steeples of the city through their binoculars. But then General Winter came to the rescue of the Russians. In early December, heavy snow fell followed, as is often the case, by bitter cold. On several nights temperatures fell to 45 degrees below zero Fahrenheit. The German high command had largely ignored the logistical needs of their armies. No provision had been made for winter clothing. Soldiers froze to death. Their tank engines would not start. The heavy snow made maneuver almost impossible in any event. Severe winter in December in northern climes is not unusual. But, it did seem to come somewhat early in 1941. Could Divine Providence have had a hand in the matter?

The Red Army of course, knew how to survive and cope with their severe winters. They were in their element. The Germans were not. General Winter did not defeat the German army, but it surely stopped them in their tracks. They were so close to their goal, but so far.

Some have questioned why the Wehrmacht did not take winter more seriously. Part of the problem was German priorities wherein logistics and things as mundane as winter clothing were a low priority over bullets and artillery shells. More on that shortly. Furthermore, Hitler's hubris and arrogance led him to believe his armies would roll over Russia long before winter set in. Moreover, winter in Berlin was nothing like winter in Moscow.

First, Moscow is at a higher latitude than Berlin, making it colder by that simple fact. Furthermore, northern Germany is on the shores of the North and the Baltic Seas. Large bodies of water always moderate winter temperatures and Germany enjoyed that moderation. Winter came, but not like in Russia. Moscow on the other hand was not near any large body of water. It was more or less in the center of the continent of Eastern Europe. Such cities are always much colder than coastal regions. Some in the Wehrmacht no doubt understood that. Apparently, Hitler did not.

Though the Germans had had smashing success in killing or capturing millions of Russian soldiers, just beneath the surface things were not so rosy. The German army had given low priority to its logistical needs. Perhaps this was because Hitler had assured them the war in Russia would be over in short order. Perhaps, they had not had to deal with major transportation issues in their forays into Poland, Denmark, or France.

Russia was a different story. It was immense. Germany had to send supplies many hundreds of miles over a road system that was dusty when dry or mud when wet. Much of the Russian rail system was of a different gauge than the German railroads. That meant for the Germans to utilize captured Russian railroads, they had to be rebuild it to accommodate German rolling stock. Moreover, Germany had never built hundreds of thousands of standardized trucks like the American two-and-one-half ton, all-wheel-drive truck. They rather had cobbled together a hodge podge of trucks of many different manufacturers which they had appropriated from the Poles, the Belgians, the French, vehicles left behind by the British at Dunkirk, not to mention vehicles from Czechoslovakia and other European countries. Spare parts were impossible. As this motley fleet of trucks tried to supply the Wehrmacht in their drive into Russia, many eventually broke down and that was all she wrote for the most of them. The Wehrmacht was reduced to using horsepower, literally, to move much of its supplies, but that produced its own issues of providing fodder for the horses.

Germany ironically did not go to a full-war footing in its economy until later in the war. Russia, by contrast, immediately moved to a full-war mobilization of its industries and economy (as did the United States when it entered the war and as did Britain). In the early years of the war, Hitler did not want to cause rationing and shortages for the civilian population in Germany. He feared a political backlash. The upshot of that decision was that as things began to get tough in the Russian campaign, German industry began to have difficulty in keeping the Wehrmacht supplied.

The farther the Germans drove into Russia, the more severe were its shortages. Whereas the Red Army fired what seem like an unlimited number of artillery shells against the Germans, the Wehrmacht in turn had to ration how many shells their units could fire against the Russians. Their logistics simply could not keep up. They had the same trouble with fuel for their tanks, not to mention food, clothing, and other equipment and supplies necessary to keep a modern army moving. The fact the Germans had alienated the Russian civilian population only made things worse. Russian partisans (underground fighters) continuously attacked German supply lines, seizing supplies for themselves and denying them to the German front-line troops.

On December 5, 1941 General Zhukov ordered his newly reinforced armies in front of Moscow to attack the Germans. They punched through the frozen Wehrmacht. German weapons were frozen, German soldiers were frozen, and sometimes soldiers were frozen to their weapons. The Germans could only watch helplessly as the Red Army, warmly clad in furlined coats, fur-lined hats, fur-lined boots, and camouflaged in white snow-suits, emerged like ghosts through the fog and snow.

Furthermore, the Russians now introduced into the battle their newly designed T-34 tank which was superior to any tank the German Panzers had at this point. In fact, all things being equal, the T-34 was probably the premiere tank of any nation in World War II. It was fast and nimble. It had sloped armor which was almost impervious to German Panzer III or IV tank fire. It utilized a pneumatic starting system which did not require start batteries that failed in bitterly cold weather—which the Germans learned to their dismay. It utilized a diesel engine whose fuel was much less flammable in battle. It had wide tracks which could maneuver and navigate through deep snow and mud compared to the narrow tracks of the German Panzer III and IV tanks, which could not. The Russians of course knew to use lighter-weight winter oils in their tanks. The German had not thought of that and their tanks became difficult if not impossible to start.

The T-34 tank was superior in every respect to the German Panzers. Hitler later reportedly said that if he had known about the T-34 tanks, he would not have invaded Russia.

Though the Germans were superior as a ground army (at least when its equipment worked), the Russians simply had more manpower. One German general would later write in his diary that when German forces destroyed a dozen Russian divisions, the Red Army simply moved up another twelve divisions to take their place. The Germans did not have a seemingly limitless supply of manpower. The Russians did. Though the Germans would slaughter literally millions of Russian troops during the Russian campaign, the Russians always had more. That would prove to be the case as the Battle for Moscow continued.

As Zhukov's forces tore into the frozen German lines, Hitler's interference and ongoing blunders only made things worse for the Wehrmacht. He had assumed full command of German forces, not only of the SS, but also of the Wehrmacht. He did not fully trust his general and often considered them timid or not aggressive enough to suit him. Furthermore, he was resentful and suspicious of most of them. They pled with him to allow German forces to make a tactical retreat to more defensible positions. Hitler feared a retreat would disintegrate into a panic-stricken rout. He ordered the Wehrmacht to hold their positions to the last man, utilizing a hedgehog defense of strong points that could be defended even when surrounded. Hitler would thereafter only listen to himself and rarely accept the advice of his generals to retreat. This meant the German armies at Stalingrad and Normandy held their positions until they were destroyed.

Zhukov pushed the Germans back to Rzhev, 150 miles west of Moscow. German lines were still intact, though battered, and their armies were still able to fight. However, Stalin now became overconfident. The Red Army also suffered terribly during their counteroffensive: many of their troops were largely inexperienced, their supply lines were strained by snow and mud, and they also suffered from the cold. Stalin ordered his exhausted forces to continue the attack. The result was major losses in hopeless attacks. By February, the Germans counterattacked, destroying several Soviet divisions.

The Russian-German war would continue to see-saw back and forth for the next two-and-one-half years. The Germans, though superior in most dynamics, did not have the strength to defeat the Red Army. The Russians and their seemingly endless supply of troops and armor, for the most part, did not have the leadership skills to drive the Germans out. The slaughterhouse that was the Eastern Front would continue into 1942, and then all the way to 1945. But in 1944-45, it was the Germans who were being slaughtered. If Hitler were to win, he had to do so before the Soviets fully mobilized, reorganized their armies, and harnessed their vast industrial potential. Hitler did not succeed in that. It took the Russian two-and-one-half years to accomplish their full mobilization and armament. Then, in the summer of 1944 they slowly, but inexorably drove the Germans back to Berlin.

CHAPTER 16

The Battle of Stalingrad

Meanwhile, after the debacle of the attempt to attack Moscow and the terrible winter of 1941-42, the Germans hunkered down into the spring of the year. They licked their wounds, tried to keep from freezing to death, and slowly regrouped. As summer came, with the prospect of major combat, the Russians had another surprise for the Wehrmacht. General Zhukov, well aware that though the Germans had been repulsed from Moscow the preceding winter, knew they were still less than 150 miles to the west near Rzhev. Battles raged through the summer across the Russian-German front. The only result was many thousands of German troops were killed along with tens of thousands of Russians.

The Russians anticipated Hitler ordering his troops to make another attempt to overrun Moscow. But the Fuhrer had other plans. He had never lost sight of his goal of seizing the Caucasus region with its oil fields and oil production facilities. He therefore ordered his Army Group South to move out to the southeast toward that objective. Along the way was the Soviet city of Stalingrad, situated along the mile-wide Volga River. (The city had formerly been named Volgograd, but had been renamed after Stalin seized power.) It was a large city, stretching almost 30 miles along the Volga River.

Hitler therefore split Army Group South with one portion still pressing on toward the Caucasus region. The other half was to take Stalingrad. The Germans did not view the city of any great strategic importance, but rather only as another city along their route of march which would have to be seized. Yet the Battle of Stalingrad would become one of the largest, longest, and bloodiest battles in history. From August 1942 through February 1943, more than two million troops fought in close combat. Nearly two million people were killed or injured in the fighting, including tens of thousands of Russian civilians. When it was over, the Battle of Stalingrad would prove to be a major turning point, with the tide of war turning in favor of Russia and the Allies.

Not only was Stalingrad an obstacle to their greater objective of seizing the Caucasus region, it also was a major Russian industrial city of 400,000 citizens. It was an important cog in the Russian armaments industry, producing T-34 tanks from its tractor factory and other important military materiel, including artillery shells. It was also an important transportation center for the Russians with the navigable Volga River at its doorstep. Hitler also wanted the city for vanity reasons. It was named after Joseph Stalin and would be a great personal victory for Hitler and good propaganda value if it fell. Stalin, of course, realized its importance to Russia for similar reasons.

Hitler appointed General Friedrich Paulus over the German Sixth Army, part of Army Group South. On August 23, 1942, under direct orders from Hitler, the Luftwaffe began an intense bombing campaign against Stalingrad. JU-87 Stukas came in at low level and systematically destroyed Russian anti-aircraft batteries which were manned largely by teenage girls, barely out of high school.

After eliminating them, higher altitude German medium bombers began to systematically destroy the city. Estimates range as high as tens of thousands of civilians died in the initial bombing campaign. However, the bombing of the city would eventually backfire on the Germans. Once their troops invaded, they discovered the streets were completely blocked from the rubble of bombed-out buildings, forcing them into street-to-street, house-to-house, and hand-to-hand combat. The Soviet government refused to evacuate the 400,000 Russian civilians in the city. They would be employed, including women, to assist the Red Army in digging trenches and other necessary tasks, eventually taking up arms themselves against the German invaders. The Russians further told the population their troops would fight all the more tenaciously knowing they were defending civilians.

The Red Army was initially able to slow the Wehrmacht's advances during a series of brutal skirmishes just north of the city. However, Soviet forces lost more than 200,000 men. Unfortunately for the Russians, they only temporarily held back the German attack. But once again, the Fuhrer's hubris and arrogance ruled the day. General Paulus told Hitler at a conference on September 8 that it would only take about ten days to take the city with another two weeks to regroup before moving on into the Caucasus region. As it turned out, the battle would rage into February of 1943, almost six months later.

Paulus directly attacked the city on September 27, 1942. Because the city had already been largely destroyed by Luftwaffe bombing, the Wehrmacht was forced to engage in close combat, fighting in burned out civilian houses, from behind mounds of destroyed building, going floor by floor in wrecked apartment buildings, and even in the sewers. German soldiers called it *rattenkrieg* or a rat's war. The decision to invade the city deprived the German commanders of their great Blitzkrieg advantages of armored mobility. The Russian commander on the scene ordered his troops to entice the Germans into channels of fire where half buried T-34 tanks, antitank guns, machine guns, and hidden troops would attack the Germans from the rubble. The Red Army sent squads of men into the sewer system armed with submachine guns and crude flamethrowers. The dust of battle was so great, it was often hard to distinguish the uniforms of German or Russian soldiers. They all were covered in a dull tan-gray dust.

Food and water supplies became virtually unavailable. Russian troops would shoot holes in rain gutters hoping to catch a few drops of water dripping therefrom. Herman Goering had promised Paulus that he would keep the Wehrmacht re-supplied by air. His JU-52 transport planes (similar to American C-47s) made a hopeless attempt, but were too few in number and many of those were shot down by the Russians. The German troops in the city were starving as well. The Russians made attempts to resupply their forces by night across the Volga River, but it was always too little and too late. Meanwhile, Russian soldiers and even officers attempted to flee across the Volga on flotsam, usually by night, to escape the hell within the city. Aware of this, Soviet commanders ordered the NKVD, the Russian secret police, to shoot on sight anyone deserting the battle. More than 13,500 executions of fleeing Russian soldiers took place, either on the spot or ordered by courts martial. Moreover, Stalin had issued order number 227 stating, "Not a step back." No ground was to be surrendered to the Germans.

Paulus continued to attack into the city and the slaughterhouse continued—on both sides. September became October. The Germans were slowly seizing the city. However, little did they realize that the Russians were setting a major trap for them. The Germans were winning the battle, but unaware they were about to lose their war—the greater strategic scheme.

Zhukov and Stavka, in watching the disaster at Stalingrad unfold, saw a great opportunity. Their plan was relatively simple. They would assemble massive forces to the northwest and southwest of Stalingrad—up to 50 miles away. As much as 60% of the Red Army would become part of what the Zhukov called Operation Uranus. The Russian commander on the scene in Stalingrad did not initially realize that the Soviet high command was using him as bait. He was not allowed to withdraw from the city, though the battle was being lost. Rather, he was to keep the German Sixth Army occupied, while Zhukov prepared for Operation Uranus. At the appointed time, the Russian northern army would sweep south and the Russian southern army would sweep north in a giant pincer movement, surrounding Paulus's Sixth Army.

On November 19, 1942, Zhukov ordered the two Russian pincers to snap shut behind the German Sixth Army. By now the Germans were aware of what Zhukov was up to, but it was too late. Though there were fierce battles, Paulus was trapped. Three-hundred-thousand men in the German Sixth Army were surrounded. Hitler refused any pleas from Paulus to withdraw. Goering again promised to resupply the Sixth Army by air, but Paulus's twenty-two divisions were doomed. They did not have the fuel or the physical strength to breakout. Goering's lumbering Ju-52 transports either were frozen on the ground or shot down in considerable numbers as they passed over heavily armed Soviet forces. Hitler refused the pleas of Paulus to surrender. The Fuhrer even promoted him to field marshal and reminded him that no German field marshal had ever surrendered. Nevertheless, on January 31, 1943, Paulus surrendered. Of the original 300,000 men in his army, only 91,000 were alive at the time of surrender. By Spring of 1943, half of those had died. The rest wound up in Russian POW camps where they either died of starvation, disease, or froze to death. Ironically, of those 91,000 only 5,000 survived until after the war and were eventually repatriated to Germany in 1955.

Upon the defeat at Stalingrad, the portion of Army Group South sent toward the Caucasus region retreated back to defensible lines.

Ironically, Stalingrad did not hold any great strategic value for either the Russians or the Germans, other than the prestige of Stalin's name for the Russians and Hitler's attempt to besmirch it. Hitler's stubbornness in refusing to allow his commanders to retreat to more defensible positions once again backfired on the Germans. The cunning Zhukov turned what seemed a great tactical defeat into a major strategic victory for the Red Army. Germany and its several small allies (Romania and Italy) suffered over 800,000 casualties, including those captured or missing. The German military was significantly weakened as a result. The Soviet Union is estimated to have lost at least 1,100,000 casualties, including at least 40,000 civilians killed in the battle. The grand total of casualties for both sides was approximately two million.

The Russians consider the Battle of Stalingrad to be one of the greatest battles of what they, to this day, call the Great Patriotic War. Other historians consider it to be the greatest battle of World War II. The battle stopped the German advance into the Soviet Union and marked the turning of the tide of the war in favor of the Allies. From this point onward, with the exception of the coming Battle of Kursk, the Germans would basically be on the defensive. They would not move any farther into the Soviet Union. In the first year and a half in Russia, the Germans had lost well over one million men. Germany could not replenish those ranks. The Russians suffered much, much more, but they were able to replenish their ranks. Moreover, they were fighting on their home soil and tenaciously defended it. Their supply lines were short. They usually had adequate equipment, weapons, food, and proper clothing. The Germans were constantly short of all of the above.

CHAPTER 17

American Lend-Lease

The American Lend-Lease policy was a program under which the United States supplied friendly nations with food, oil, and materiel between 1941 and August 1945. Prior to America entering World War II, the United States determined to help those nations struggling against Nazi Germany. Though technically a violation of its neutrality in the conflict, America foresaw England, Russia, and other nations as falling to Hitler's aggression unless they were helped. In return, England in particular agreed to give long-term leases of British army and naval bases in the western hemisphere to the United States. After the United States entered the war at the end of 1941, it continued to aid its allies.

The Lend-Lease program was a testament to American largesse—much of what was "lent" to its allies was never repaid. However, the United States foresaw that Hitler would prevail unless aide was sent to friendly nations. At the beginning of 1942, the talk of world conquest by the Nazis was a very real fear. Lend-Lease was also a testament to the industrial strength of the United States in being able to fight its own war with Japan and Germany, as well as providing the necessary war materiel to its allies for them to prevail.

In regard to the Russian-German conflict, there is little question that Germany would have been victorious over the USSR if the United States had not sent massive aide to Russia. However, the motives of the United States were not altruistic. If the Soviet Union fell to Hitler, it would have been exponentially more difficult for America and England to have defeated Germany. It may even have been impossible, short of nuclear weapons which were not in the picture in 1941. Therefore, it was actually in the self-interest of the United States to give aide to not only Britain, but the Soviet Union as well. The latter was an unsavory ally, but it was a crucial ally against Hitler nevertheless. The overriding motive was defeating Hitler. Aiding the Soviets was viewed as a necessary evil toward that end.

At this point, we will only focus upon American aide to the USSR. It was huge. About eleven billion dollars in aid went to the Russians during the war. In 2020 dollars, that would approximate more than 120 billion dollars. That was second only to the aide sent to England which was about 32 billion dollars (or about 350 billion in current dollars).

As Russia found itself in an existential battle against Nazi Germany, its young men were diverted from the fields to the military or armament industries. Moreover, between the Nazis looting Russian food stocks, or burning crops (some of which was also done by retreating Red Army units to prevent them falling into German hands), Russia faced severe food shortages in 1942. Hence, one critical area of aide was in food shipments. In a nation still using draft horses, the USSR had lost about 60% of its draft animals to the invading Germans. Tens of thousands of pieces of farm equipment from tractors to harvesting machines were either destroyed as the Germans moved deeper into Russia or were simply appropriated by them for their own purposes.

Lend-Lease thus provided a massive amount of foodstuffs and agricultural products.

As the Russian economy, almost overnight, shifted from domestic production to war armaments, production of railroad equipment—locomotives, rails, and rolling stock was shunted aside for the production of tanks and other heavy weapons. Therefore, the United States sent 1,922 locomotives to Russia along with over 11,000 railroad cars. As the war accelerated from 1941, so did American shipments which reached a peak in 1944.

Among other goods, LendLease supplied:

58% of the USSR's high octane aviation fuel,
33% of their motor vehicles,
30% of fighters and bombers—18,200 fighters and bombers in all,
93% of railway equipment (locomotives, freight cars, wide gauge rails, etc.),
50-80% of rolled steel, cable, lead, and aluminum;
43% of garage facilities (building materials & blueprints),
12% of tanks and self-propelled guns,
50% of TNT and 33% of ammunition powder (in 1944)
16% of all explosives.

The Russian military focused on things such as tanks, artillery, and military aircraft. However, they made little provision for utilitarian items such as military trucks during the war. The United States sent them over 400,000 trucks, mostly the renowned deuce-and-a-half (two-and-one-half ton) trucks so ubiquitous to the US Army. Included in that number were also thousands of jeeps. By the end of the war, about one third of the Russian truck fleet had been made in the USA.

These massive amounts of aide were shipped by ocean freighters to major ports in Asia. The largest amount was 8,244,000 tons delivered to Siberian ports on the eastern Pacific coast of the USSR. It was then transshipped by the Trans-Siberian railroad westward. Another almost 4,000,000 tons arrived at Archangel and Murmansk in northwestern Russian on the Arctic Ocean. Over 4,000,000 tons of aide arrived at ports of Iran and then were transshipped north by rail into Russia. About 700,000 tons were able to be shipped through the Black Sea, late in the war, to Odessa and Sevastopol in Crimea.

The risk of submarine attack was always present, particularly in 1941-42 when many American ships were sunk. During the war, 632 American merchant ships were sunk by German U-Boats. From February to May of 1941 alone, 348 allied ships were sunk. But overall, 97% of the aide shipped arrived safely in the Soviet Union.

In retrospect, the USSR provided the manpower, but the United States supplied a significant amount of war materiel, without which, the Soviet Union likely would have been conquered by Hitler. A Russian historian by the name of Boris Sokolov later said,

> "On the whole the following conclusion can be drawn: that without these Western shipments under Lend-Lease the Soviet Union not only would not have been able to win the Great Patriotic War, it would not have been able even to oppose the German invaders, since it could not itself produce sufficient quantities of arms and military equipment or adequate supplies of fuel and ammunition."

After the war, Nikita Khrushchev remarked about private conversations he had had with Stalin.

> "He (Stalin) stated bluntly that if the United States had not helped us, we would not have won the war. If we had to fight Nazi Germany one on one, we could not have stood up against Germany's pressure, and we would have lost the war."

In 1963, Soviet Marshal Georgy Zhukov said,

> "Today some say the Allies didn't really help us . . . But listen, one cannot deny that the Americans shipped over to us materiel without which we could not have equipped our armies held in reserve or been able to continue the war."

Other Soviet leaders have tried to downplay American aide. But Soviet leaders including Stalin and Zhukov understood. American industrial might not only helped win the war for the American armed forces, but also for its allies in Britain and Russia.

CHAPTER 18

The Battle of Kursk

After the debacle of Stalingrad, the Wehrmacht worked though the late winter and spring season of 1943 regrouping, refitting, and planning for summer operations. The German-Russian front still stretched from Leningrad to the Black Sea. But by the Spring of 1943, the Russians had pushed a massive salient—a protrusion, or bulge—into the German line. It was approximately 150 miles from north to south and 100 miles from east to west. At the center of the salient lay the Russian city of Kursk. Among other things, Kursk was a regional hub for Russian railroads and transportation. Hitler and his general staff saw an inviting opportunity. They debated and argued over plans through the Spring season, but the prospect of pinching off a Russian salient with a massive pincer movement was irresistible to Hitler.

The strategy was simple enough. The German Ninth Army would mass for an attack at the northern base of the Russian salient and drive southward, while the Fourth Panzer Army would likewise attack northward into the salient from the south. Upon meeting, the giant pincer movement would trap a major element of the Red Army and lead to its destruction.

Thus, the Germans began to gather forces and weapons. The northern German Ninth Army massed 300,000 men on the northern flank of the Russian salient. The southern German Fourth Panzer army amassed another 300,000 troops on the southern flank. The Wehrmacht also assembled over 2,700 tanks for the upcoming battle. Hitler for the first time introduced the brand new Panther tank which was designed to be superior to the very effective Russian T-34 tank. However, the Panthers were rushed to the front and had not been thoroughly field tested. Consequently, in the coming battle, they suffered significant reliability problems and break downs.

The Russians were aware of the German build up. Moreover, they were specifically aware of their plans. The British had broken the German military radio code and therefore were intercepting in detail the German plans for the coming attack. They passed that information on to the Soviets. Zhukov therefore ordered massive defensive preparations. The Russian side of the salient soon bristled with tank traps, hundreds of miles of trenches, with multiple belts of defensive fortifications. Should the Germans break through one, they faced another belt of Russian defense, and then another. Zhukov amassed 1,300,000 men in the salient along with over 3,600 tanks. He further held another 500,000 troops in reserve along with an additional 1,500 reserve tanks.

The German code name for the attack was Operation Citadel. Hitler had planned to attack on May 3, 1943. But once again, his generals were reluctant, aware of the massive Russian build up. Hitler therefore decided to delay the attack ostensibly over bad weather and because he was awaiting the full arrival of his new Panther tanks.

Zhukov took advantage of Hitler's delay and further fortified his lines with more tank traps, barbed wire, and close to one million anti-personnel and anti-tank mines. The genius of previous German attacks had been their Blitzkrieg tactics of surprise and fast mobile attack. With all the delay, the German's lost the advantage of surprise. The Soviets knew the Germans were coming and had ample time to prepare.

Early July 3, 1943, the Red Army ordered a pre-emptive artillery barrage hoping to soften up the impending German attack. It only slowed the Wehrmacht slightly. The Germans whereupon launched their main attack with their own heavy artillery barrage, followed by Stuka attacks, and then their armored units moving out. The Soviet defenses withstood the German onslaught. Though the battle raged for another week, the Red Army stopped the northern German force. On July 12, the Russians launched a counterattack north of Kursk. They breached the German lines and by July 24th had the Germans on the run, pushing them back beyond Operation Citadel's original starting line.

To the south, the Wehrmacht slogged its way northward with some success, arriving at the Russian village of Prokhorovka, about 50 miles southeast of Kursk. There the largest tank battle in history (to this day) erupted. On July 12, tanks of Russia's Fifth Guards Tank Army violently collided with the tanks of Germany's II SS Panzer Corps. As the Red Army units put pressure on the Panzer Corps, the largest armored clash of World War II erupted west of Prokhorovka. A Russian officer described the developing scene, "A cyclone of fire unleashed by our artillery and rocket launchers . . . swept the entire front of the German defenses." As Soviet artillery ceased firing, the coded order, "Stal! Stal! Stal!" (Steel! Steel! Steel!), was issued and the Russian tanks went into action. Within minutes another great cloud of dust billowed skyward as Soviet T34s and T70s picked up speed.

At the sight of the dust cloud, the Germans commanders issued warning to front-line units to prepare for the imminent arrival of Soviet tank formations. Each panzer tank commander followed a well-practiced routine: he stopped his tank, identified a target, the gunner lined up the victim in his sights and then, when ordered, opened fire. The process took just a few seconds, and was as efficient as it was effective. A German officer would later relate, "What I saw left me speechless. From beyond the shallow rise about 150-200 meters in front of me appeared fifteen, then thirty, then forty tanks. Finally there were too many to count. The T34s were rolling forward toward us at high speed, carrying mounted infantry standing on the engine compartment and clinging onto handles welded onto the hull. . ... Soon our first round was on its way and with its impact, the T34 began to burn."

A Russian tank commander later related, "The distance between the tanks was below 100 meters—it was impossible to maneuver a tank, one could just jerk it back and forth a bit. It wasn't a battle, it was a slaughterhouse of tanks. We crawled back and forth and fired. Everything was burning. An indescribable stench hung in the air over the battlefield. Everything was enveloped in smoke, dust and fire, so it looked as if it was twilight . . . Tanks were burning, trucks were burning."

Commanders on either side found it difficult to tell friend from foe, as dust and heavy black smoke combined with darting Russian tanks to produce a deadly chaos. Because of their superior training, German Panzer units excelled in such difficult situations. Their crews worked together to outthink and outfight their more numerous Russian opponents. Clear thinking, previous experience, two-way radios in their tanks, and superior tactics allowed the Germans to destroy six times more Russian tanks than they themselves lost. The result was a German tactical victory at Prokhorovka—a remarkable achievement considering their numerical inferiority.

Though the Germans had achieved a tactical victory at Prokhorovka, they lost the greater strategic battle. The Germans could not snap their pincers shut, in fact, they did not even come close to so doing. They could not breach Soviet defenses. Hitler blundered again. His grandiose plans, once again blew up in his face. The Russians suffered huge losses but still managed to prevent the Germans from breaching their defensive belts, which effectively ended the German offensive. And with that defeat, thus ended the last major German offensive in Russia in World War II.

Years later German Field Marshall Eric von Manstein would write, "The last German offensive in the east ended in fiasco, even though the enemy (i.e., the Russians) opposite the two attacking armies . . . had suffered four times their losses in prisoners, dead and wounded." The Red Army won but with devastating losses. Despite numerical superiority over the Germans in manpower and tanks, they suffered four times as many casualties. Historians estimate there were up to 800,000 Soviet casualties compared to 200,000 for the Germans. Other historians believe those numbers are lower than the reality. The Russians lost over 1,600 tanks and fighting vehicles. In contrast, the Germans lost 252 tanks. It was the basic story of the eastern front. The Germans were superior in their leadership, tactics, weapons, and discipline. But they simply could not overcome the virtually endless supply of cannon fodder of troops which the Soviet high command fed into the meat grinder of battle. Soviet brute force simply overcame superior German skill in operations.

As the Battle of Kursk raged, Allied forces landed troops in Sicily on July 10th. That diverted Hitler's attention and he began moving forces from Russia to Italy to deal with the expected Allied invasion there. Hitler never regained momentum on the Russian Front and never recovered from his loss of manpower and armor. The Germans were forced onto the defensive and never again would launch another major offensive in Russia.

Late 1943 into Spring 1944

After the fiasco at Kursk, the Germans had lost well over one million men in Russia. Russia could replenish such losses, Hitler could not. He was reduced to drafting men up to 50 years old as well as teenagers from the Hitler Youth organization. Though there were no more titanic battles like Stalingrad or Kursk, the Red Army continued to press the Germans. In September of 1943, the Russians pushed the Wehrmacht back across the Dnieper River. By the mid-winter of 1944, they had retaken all of the Ukraine region. Further north, in early January 1944, the Germans had been pushed back to the old Polish-Russian border. By the end of January 1944, the Russians had broken the siege of Leningrad. Through the late winter and coming spring season, the Russians began positioning themselves for the coming attack against Germany proper.

Hitler's invasion of Russia had come to an end. After almost three years of brutality, atrocities, and millions dead, Operation Barbarossa was a complete failure. Though the Wehrmacht was initially successful, the resurgent Red Army—with help from American Lend-Lease—had driven Hitler's forces from Russian soil. In the coming year, they along with the Allies would destroy Germany altogether.

Hitler had bit off more than he could chew. His hubris, pride, and arrogance clouded his judgment. His deranged obsession of annihilating the Jews further led to one blunder after another.

His greed and covetousness for more *lebensraum* backfired completely. He had underestimated the Russians in virtually every dynamic, thus violating one of the most basic axioms of warfare—never underestimate your enemy. He underestimated the Russian determination to fight to defend their homeland. He underestimated their industrial potential to produce war materiel. He underestimated the ability of Stalin to call up millions of replacement troops after the initial debacle of the German invasion. He underestimated the Russian weather. He underestimated the vast expanse of the country he was trying to conquer. He underestimated the problems of keeping his armies supplied in the expanses of Russia.

Though Hitler made various tactical blunders all along the way, his major, catastrophic blunder was invading Russia in the first place. Had he been content to rule over Western Europe from the Russian border to the Atlantic coast, as he did in the spring of 1941, things would have turned out much differently. Western Europe might still be part of a German Reich today.

* * * * *

CHAPTER 19

Hitler's Second Catastrophic Blunder

On December 7, 1941 Japan attacked Pearl Harbor. As Franklin Roosevelt announced the next day, it was a day which would live in infamy. The United States thereupon declared war against Japan on December 8th. On December 10, 1941, Hitler (i.e., Germany) declared war on the United States.

In so doing, Hitler committed his second *catastrophic* blunder. This coupled with his invasion of Russia about six months earlier categorically ensured his defeat. For all its advancements, particularly in military matters, Germany simply could not prevail against the combined might of the United States, Great Britain, and the Soviet Union. Words such as stupid, dumb, idiotic, foolhardy, and others come to mind. As Hitler had grossly underestimated the Soviet Union, even more so he underestimated the might of the United States. Nevertheless, his hubris, arrogance, and ideology overruled the scope of any good judgment.

The United States during the 1930s was consumed in dealing with the Great Depression. The nation had been reluctant to get

involved in World War I, but it did and of course was the catalyst for victory over imperial Germany. However, throughout
the 1920s and 1930s Americans wanted nothing to do with more
European wars. The nation had become strongly isolationist.
Even Franklin Roosevelt in running for his third term as president had promised voters he would not get America involved in
another war in Europe. Yet, as England found itself at war with
Germany in 1939, Roosevelt was determined to help the English,
to the point of violating international norms of neutrality.

Throughout 1940-41, the United States, at first surreptitiously
and then rather openly, was shipping critical supplies to England.
Foreseeing war coming to America, Roosevelt had requested and
Congress grudgingly authorized a peace-time draft. The country began to rearm itself. Through the Lend-Lease Act of March
11, 1941, the United States officially began sending war materiel
to England, China, and then later to Russia. United States forces
were sent to replace British forces in Iceland. In early September
1941, a German submarine had fired on the American destroyer
USS Greer and missed. Roosevelt thereupon issued a "shoot on
sight" order on September 11, 1941 against all German warships.
This for all intents and purposes unofficially declared a naval war
on Germany and the looming Battle of the Atlantic.

Ironically, Hitler had no means to defeat the US. And, he really had no means to even effectively attack the United States.
Germany was a land power. In 1941, Germany had no blue-water navy. Because of its losses in World War I and because of
treaty limitations after the war, Germany basically had a coastal-
defense navy. The Kriegsmarine (German navy) had four relatively new pocket (i.e., small) battleships and a number of old
ones, along with seven battle cruisers. The Kriegsmarine also
had a number of light cruisers and destroyers. But it had no
aircraft carriers or heavy battleships. Its military threat against

Britain or the United States Navy was basically in attacking cargo ships. The Kriegsmarine was no match for either the British or American navies. Hitler therefore had no means of reaching American shores with anything more than a token force. (Of course, German U-boats wreaked havoc on American shipping early in the war.) He also had a grandiose idea called Plan Z in which Germany would build a first-class, blue-water navy able to defeat either the American or British navies. His goal was an 800-ship navy by 1947. Of course, that never happened, or even began to happen. As the great land battles in Europe erupted, Plan Z was shelved.

Therefore, come December 10, 1941, Hitler still did not have a blue-water navy, yet he declared war on the United States. He had no means of transporting troops across the Atlantic Ocean even if he wanted to. In 1940, he could not even manage to transport invasion troops to England from France, 20 to 50 miles away.

Furthermore, Hitler had no long-range, strategic air force. The German Luftwaffe was essentially an infantry-support, tactical air force. Its medium bombers barely had enough range to bomb England from the continent and back again. Its fighters had even less range. For infantry support, the Luftwaffe fielded an excellent dive bomber in the JU-87 Stuka. But it was no long-range bomber. Germany had no four-engine, heavy, strategic bombers with which it could bomb the United States. And even if he had them, Hitler would have needed a base in the western hemisphere to launch them as the United States later would do from England. Hitler had none of the above.

Other than harassment in sinking American cargo ships, Hitler had no means to inflict military damage upon the United States. Yet he declared war on America.

What is even more astonishing is that Hitler declared war on America at the same time His armies in Russia had bogged down and things were beginning to sour for the Wehrmacht there. Hitler never wanted a two front war, but by declaring war on America, he surely got one.

Once again, his hubris, arrogance, and pride led him to make a catastrophic blunder. Hitler certainly was an intelligent man. That was evident as he led Germany out of its depression to being a world power during the 1930s. Yet, intelligent people sometimes make stupid decisions. Hitler was the classic example thereof.

The Fuhrer declared war on a country whose population was almost twice that of Germany and whose industrial capacity for making weapons of war was many times greater than Germany. He declared war on a country which had sufficient oil supplies in distinction to Germany which had to import its oil. Though America's army in 1940 was weak, Hitler grossly underestimated the United States, to his own peril.

Why Did Hitler Declare War on America?

Historians have debated why Hitler declared war on the United States. Various reasons have been postulated. One thought is that he was allied to Japan and America had declared war on Japan on December 8, 1941. On September 27, 1940, Germany had signed the Tripartite Pact, also known as the Berlin Pact, which was an agreement between Germany, Italy, and Japan. Article 3 of the pact stated that any of the three signatories would "undertake to assist one another with all political, economic and military means if one of the Contracting Powers is attacked by a Power at present not involved in the European War or in the JapaneseChinese conflict." Reading between the lines, who they had in mind was the United States.

Yet, the German and Japanese alliance was only one of convenience. Neither helped the other in any significant way during World War II. Any cooperation was only superficial. Neither Germany nor Japan had any common goals even after Pearl Harbor, other than defeating the United States.

More likely, Hitler declared war on the United States because the US was already supplying war materiel to England through the Lend-Lease Act. His war declaration likely was to spite the United States for their aide to England. Moreover, in declaring war on America, what had been moderate attacks on American shipments to Britain, suddenly became a torrent. German U-boats appeared off the Atlantic and Gulf coasts of the United States. In 1942, they unleashed a salvo of attacks on American oil tankers, cargo ships, and military vessels, often in sight of American coastal cities. Hitler's objective was twofold: (1) to diminish shipments of war material destined for England; and (2) to spite America for helping England. Declaring war on the United States gave Hitler the legal basis to attack American shipping on a wholesale basis. Hundreds of American ships were sunk by German U-boats in 1942-43.

But Hitler chose an inopportune time to declare war on the United States. Through the 1930s until December of 1941, the United States was largely isolationist, wanting no part in Europe's wars or Japanese aggression in Asia. The Republican Party was largely isolationist, but after Pearl Harbor, the United States overnight sought revenge on Japan. The Congress almost unanimously voted to declare war on Japan. (The vote was 82-0 in the Senate and 388-1 in the House.) Because Germany was viewed as Japan's ally and because Germany declared war on America December 10, 1941, the spirit of retribution against Japan also came down against Germany in the American public. Admiral Yamamoto, who commanded the attack against Pearl Harbor, remarked not long thereafter that he feared all Japan had done was awaken a sleeping giant and filled it with terrible resolve. He was absolutely correct and part of that resolve became focused against Nazi Germany. Hitler's timing could not have been worse—another blunder.

As Hitler had underestimated Russia as an enemy, so He grossly underestimated the United States. Once again, his arrogance and hubris led him to make a disastrous blunder.

The United States had a weak army in the 1930s. Based upon the number of men in uniform, the US Army was the 16th strongest army in the world, just ahead of Romania. There were only 160,000 men in the entire US army and they were woefully ill-equipped. Moreover, there was poor morale and low discipline. Soldiers trained with wooden guns. American tank forces trained with cardboard tanks placed over pickup trucks.

But the US Army was being rebuilt by the Roosevelt administration. The same was true for the army air force. The ubiquitous B-17 flying fortress heavy bomber was designed, developed, and tested during the late 1930s. It would of course become the workhorse of the Unites States Army Air Force in bombing Germany. Other American bombers such as the B-24 Liberator heavy bomber were designed and initially built before America entered the war. Likewise, the B-25 Mitchell medium bomber was brought online before America entered the war. War was looming and America was slowly preparing itself. Hitler saw the weak American Army of the 1930s, but he was oblivious to the rapid re-armament and reconstituting of the United States armed forces in 1940-41.

The most basic defense America has had since its inception is the great oceans on each side of the country. That remains true to this day. The United States therefore already had a powerful Navy even before the war. It was the second most powerful navy in the world in 1941—England being first. The US Navy, however, would grow exponentially over the war years to become the most powerful Navy the world has ever known. By war's end in 1945, the United States Navy had added nearly 1,200 major combatant ships, including twenty-seven major aircraft carriers, eight modern "fast" battleships, and ten prewar "old" battleships. Various sources list differing numbers, but conservatively, the United States accumulated at least 105 aircraft carriers of all types in World War II. Sixty-four of them were of the smaller escort carrier type. Yet, Hitler had no clue of whom he was going up against.

Merchant Marine

Once the war got underway, the United States set up a Maritime Commission which oversaw the construction of general cargo merchant ships. Techniques were developed to mass produce them, led by American industrialist Henry Kaiser. Two basic classes of ships were built: (1) the Liberty Ships and (2) the Victory Ships. Over 2,700 Liberty Ships were built during the war along with over 500 Victory Ships. Assembly-line American building procedures allowed US shipyards to turn out an average of three ships every two days. Their design was simple and every one of them was basically the same. But the Liberty Ship came to symbolize US wartime industrial output. Though Hitler tried his hardest to sink American shipping, particularly in the early months of the war with the US, the simple fact is that American know-how and industrial might produced more ships faster than Hitler could sink them. As it turned out, the German U-boat force sank a total of 632 American ships during the war. The United States produced over 3,200 Liberty and Victory Ships during that time. Once again, Hitler profoundly underestimated his foe.

By the end of World War II, the United States had over 12 million men (and a small number of women) in its armed forces. The total for the war was over 16 million—approximately 11,200,000 in the Army, 4,200,000 in the Navy, and 660,000 in the Marine Corps. The combined Allied armed forces may have been as high as 30 million. Altogether, Germany fielded about 13.6 million soldiers, sailors, or airmen. Hitler underestimated the enemies he accumulated during the war.

The United States produced over 300,000 military aircraft by war's end. Over 50,000 Sherman tanks, not including other models, were built during the war. Over 800,000 trucks were manufactured for the US Army and 640,000 jeeps were produced. Hitler underestimated all of that.

Though the United States did engage the Germans in North Africa, Sicily and Italy early in the war, it took the US about two-and-one-half years after Pearl Harbor to fully mobilize, organize, and ramp up its military production. But by June of 1944, the US with its British allies invaded France, famously known as D-Day.

Hitler was then in a vise from which he could not extricate himself. Some allied generals were organizers and plodders such as the British Bernard Montgomery. Others were aggressive in taking the battle to the Germans such as George Patton. But by fits and spurts, the allied forces quickly began to close the battle against Germany itself. Meanwhile, by the summer of 1944, the Red Army had fully mobilized, with over 500 divisions, and was methodically grinding the Wehrmacht asunder. In less than a year, the allies coming from the west and the Red Army coming from the east crushed Germany in a gigantic military vise.

When the war was over, Germany was not only defeated, it had been utterly destroyed. If Hitler had not blundered in attacking

Russia in June of 1941 and then declaring war on the United States in December of that same year, Europe today might still be part of a German empire—Hitler's Third Reich. [Hitler defined the Holy Roman Empire (800-1806) as the "First Reich," and the German Empire (1871-1918) as the "Second Reich."]

CHAPTER 20

America's Three Phase War Against Germany

The Roosevelt Administration made a strategic decision very early in the war to give priority to the war in Europe, and specifically to prevent England from falling to the Nazis. In reality, a great deal of military resources were poured into the Pacific war, particularly through the US Navy. But America's stated priority was to win the war in Europe first. That would involve three distinct phases: (1) winning the Battle of the Atlantic; (2) winning the air war over Europe; and (3) finally winning the ground war. Not much could be accomplished until the US and Royal Navies neutralized the German U-boat menace. Then, large air force materiel and personnel could be transported to England. Though American heavy bombers could manage to fly across the Atlantic, the necessary medium and tactical fighter bombers did not have the range to do so. They had to be shipped by water, hence, the importance of securing the sea lanes. An invasion landing in France would never succeed unless the Allies had air superiority over northwestern Europe. And so, the war in Europe, from the American perspective, meant first neutralizing the U-boat menace, then establishing air superiority over Western Europe, followed by the massive invasion later known as D-Day.

CHAPTER 21

The Battle of the Atlantic

B y the time the United States entered the war at the end of 1941, England had been at war with Germany for more than two years. From 1939 until the end of the war, 3,500 merchant ships were sunk mostly by German U-boats. (U-boat is an anglicized version of the German word UBoot, a shortening of Unterseeboot.) The number sunk was mostly from British or Commonwealth nations with 632 American ships sunk after the US entered the war. At least 35,000 merchant seamen perished as their ships were sunk from beneath them. The years 1942-43 were especially grim for Allied seamen.

As an island, England's existence as a nation depended upon the importation of 50% of its food, 100% of its oil, and other strategic commodities such as iron ore. This materiel came from its commonwealth and dominion nations: Canada, India, South Africa, Australia, New Zealand, and others. The United States had been discreetly sending aide by merchantmen since early in the war. Then the US shifted into high gear with the Lend-Lease program begun in March 1941. As the war developed, Britain needed more than a million tons of imported commodities *per week* in order to survive and fight. The Allies struggled to supply Britain and the Germans tried their best to stop it. Eventually, Germany lost the

Battle of the Atlantic, but Germany's loss was also at great cost to the Western Allies. It has been called the "longest, largest, and most complex" naval battle in history. It raged until the end of the war, though by 1943, Germany had largely lost the fight.

Winston Churchill would later write, "The only thing that really frightened me during the war was the U-boat peril. I was even more anxious about this battle than I had been about the glorious air fight called the Battle of Britain." Early in the war, the Kriegsmarine utilized surface ships ranging from capital ships to camouflaged, armed merchantmen to attack British convoys. They proved highly effective, sinking thousands of tons of British shipping. However, once the Royal Navy began sinking German capital ships, the Bismarck in particular, Hitler pulled them back to safe ports in Germany or Norway. By the time the United States entered the battle, the threat was almost exclusively from German submarines—U-boats.

German surface warships in the main were no match for the Royal Navy. Though the Kriegsmarine did sink the British battle cruiser Hood in May 1941 by the German battle cruisers Bismark and Prinz Eugen, that was the high water mark for the German navy in World War II. The British shortly tracked down the Bismark and proceeded to sink it. Another German heavy battle cruiser—the Tirpitz—did breakout into the North Atlantic, only to hide in a fjord in Norway where eventually the RAF located it, bombed it, and sank it in November 1944. Hitler thereafter kept what was left of his navy in port lest the British navy destroy it as well. German surface battleships contributed only a little to the Battle for the Atlantic.

Hitler appointed Erich Raeder as the Kriegsmarine's commander in chief. Raeder predicted that Britain could be brought to its knees by interdicting the flow of commerce to the island nation by sea. In principle he was right and Germany came perilously close to achieving that objective. Raeder in turn appointed Karl Doenitz (pronounced *dernitz*) as the commander of the U-boat campaign. He would in turn micro manage virtually all German U-boat operations by radio from his headquarters in Germany.

The British had quickly learned, dating back to sea battles of World War I, that merchant ships could best be protected by forming them into convoys as they steamed, primarily from North America, to British ports. Doenitz's strategy was a system which became known as the *wolf pack*. In it, U-boats would spread out in a long line across the projected course of a convoy. Upon sighting a target, they would rendezvous to attack as a group and overwhelm any escorting warships. Doenitz's plan was that as armed escort vessels pursued individual submarines, the rest of the "pack" would be able to attack the convoy with impunity. His strategy was deadly effective.

After the fall of Norway and then France in 1940, Germany thence had direct access to the North Atlantic. They accordingly built virtually impregnable submarine pens, particularly along the French coast.

June 1940 until February 1941 came to known as the "happy time" for German submariners. U-boat operations from the French bases were spectacularly successful. This eight-month period was a heyday for sinking British-bound ships. German U-boat captains perfected their Wolf Pack tactics, coordinating their attacks by radio. When one U-boat sighted a convoy, it would report it to Doenitz's headquarters, and then shadow the convoy, continuing to report as needed until other boats arrived, often at night. Instead of being faced by a single submarine, the convoy escorts then had to cope with groups of up to half a dozen U-boats attacking simultaneously. The battles reached peak effectiveness in the fall of 1940 when some convoys lost as much as 59% of their ships. Even with strong escorts of destroyers and corvettes, some convoys still lost over 25% of their ships. By the end of 1940, the situation was becoming critical for England. The "happy time" for German submariners however would begin to wane in 1941 as the British refined their tactics and new technologies were utilized.

English scientists had developed a new technology known as ASDIC. The term was an acronym for "Allied Submarine Detection Investigation Committee." The Americans would later call it SONAR (Sound Navigation and Ranging). In either case, this new technology used sound waves transmitted under water to detect objects beneath the surface, namely submarines. Though the early ASDIC sets were crude by modern standards and had their limitations, they would prove to be a great aide to detecting German U-boats. ASDIC was not decisive in the Battle of the Atlantic, but it became one of a number of technological advances that helped eventually win the battle for the Allies.

With the entrance of the United States into the war in December 1941, Germany was offered a vast new area of operations. The U-boat offensive off the US East Coast in early 1942 found American shipping completely unguarded. For all its might, the US Navy had no plans and had made no preparations for anti-submarine warfare, particularly at its doorstep. American merchant shipping losses soared between January and June 1942. More tonnage was lost off the US coast than the Allies had lost during the previous two and one-half years combined. The Germans sank nearly 500 unescorted ships off the US east coast in the first half of 1942.

The United States Navy in early 1942 rather was focused on taking the war to Japan. It was a major combatant navy and was not interested in such mundane chores as shepherding merchant shipping. Therefore in early 1942, American anti-submarine tactics relied upon private yachts and civilian coastal watchers to spot German submarines near American shores. As American shipping ramped up to transport needed commodities to England, captains simply set sail for England hoping for the best. They became sitting ducks for lurking U-boats. The American naval high command took the view that "we will do it our way" and was not interested in the advice from the British. The results were disastrous.

The British had developed an effective system of shepherding ships across the Atlantic in convoys. But convoys had their negatives. (1) A convoy was reduced to the speed of the slowest ship in the convoy. And, some of the old tubs being used were indeed slow. (2) It took time and discipline to organize each convoy. Moreover, (3) someone had to be in charge. Some American captains were not interested in the time spent in assembling and organizing each convoy or being under someone else's authority. They set out on their own at their own peril. Finally, the US Navy had to admit the British had learned the hard way that convoys were the best option for moving cargo across the Atlantic. American leaders had to swallow their pride and admit the British knew what they were doing.

By mid-1942, the Americans were forming all shipping to England (and soon to Russia) in large convoys. Convoys were formed into several columns of ships, with up to five ships in each column, forming a big box of up to 60 ships. Losses were immediately reduced, though the North Atlantic remained a very dangerous place. United States Navy or Coast Guard escort vessels were assigned to the flanks of each convoy. Attacking U-boats were then fired on by the escorting warships. In September of 1942, many convoys were formed off the New York City area as the Canadian Navy added its resources to help escort the convoys.

The United States began providing long-range patrol aircraft in mid-1942 and the British did the same from their side of the Atlantic. American PBY Catalina flying boats ranged out and ahead of convoys searching for German submarines. The U-boats tended to travel on the surface for a variety of reasons, submerging only when nearing targets or facing attack themselves. The British used converted American B-24 Liberator bombers as reconnaissance and patrol planes. Upon spotting German subs, the aircraft overhead would radio the alarm and if friendly warships were near, guide them in for an attack on the U-boats. In some cases, the big lumbering aircraft would go in for an attack on their own with modest bombs or machine gun fire. That was not their purpose or design, but they certainly kept Germans submarines off balance, forcing them to submerge and hide.

As helpful as the long-range patrol aircraft were, they could not cover the entire voyage from North America to Britain or later Russia. Both the PBY Catalinas and B-24 patrol planes could only reach about one third of the way before having to return to their bases. This left a substantial gap in the mid-Atlantic where there was no air cover. The Germans quickly figured that out and thence focused their attacks in mid ocean. This would prove to be a serious problem, until escort carriers were built.

In mid-1942, the US Navy (as well as the Royal Navy) began building and deploying escort aircraft carriers. These were sometimes called "jeep carriers" or "baby flattops." They would typically be half the length and a third the displacement of the big, fast, fleet aircraft carriers. They tended to be slower, carried fewer planes, and were less lightly armed. Most escort carriers were built upon commercial ship hulls. They thus were cheaper and could be built quickly. In the Battle of the Atlantic, escort carriers were used to protect convoys against U-boats, particularly in the mid-Atlantic gap. Fighter bombers flying off the decks of the small carriers kept constant patrol above and often would swoop down to attack U-boats they caught on the surface. Initially escort carriers accompanied the merchant ships to fend off attacks from enemy aircraft and submarines. However, as numbers increased later in the war, escort carriers also formed part of hunter-killer groups that sought out submarines instead of working with a particular convoy. As the Battle of the Atlantic progressed, the hunters were becoming the hunted.

Another technological innovation contributed greatly to the allied cause. In 1935, British scientists developed what came to be known as *radar*. The latter stands for "radio detection and ranging." It would become one of the game-changing technologies of World War II. Initially, it was used by the British for air defense during the Battle of Britain, with a series of large antenna towers placed along the southern coast of England. As its name implies, radar could detect incoming German bombers and estimate their range and direction. However, the early air defense radar systems used relatively low frequency radio pulses which required huge antennas. However in 1940, British scientists developed a ten-centimetric wave length (or micro wave) radar utilizing what was called a cavity magnetron. The significance of this innovation was that much, much smaller antennas could be mounted on ships and even aircraft.

Britain quietly collaborated with Raytheon Laboratories in the United States to produce effective radar units for military use. One such use was the mounting of radar units on both US and Royal Navy ships patrolling the North Atlantic as convoy escorts. German U-boats preferred to travel on the surface because (1) they were much faster there than submerged; (2) the could run on their diesel engines rather than their batteries; (3) the often fetid interiors of the subs could be aired out for the sake of the crews. In any seaway, their low profile at any distance made them difficult to see visually; but radar could—even just a periscope. Moreover, the preferred time for attacking convoys was at night. U-boats were thus invisible visually, but once again were quite noticeable on radar. Escort vessels and patrol aircraft could then focus firepower on the lurking sub. Radar became a great force multiplier for allied escort vessels and aircraft. Unless German subs remained submerged, they became good targets.

As the convoy system was refined, particularly on the undisciplined American side of the Atlantic, convoys became less and less vulnerable. Often a US Coast Guard officer was assigned to be the flotilla commander of each convoy. He had the authority and ability to coordinate and navigate his convoy through dangerous waters. Each of these improvements tilted the battle in favor of the Allies.

Admiral Doenitz's modus operandi was for his fleet of U-boats to radio headquarters with all pertinent data regarding their location and particularly as convoys were spotted. This communication of course was all encoded through an ingenious invention called *Enigma*. Germany developed these machines which could encode and decode their military radio traffic. It was employed extensively by Nazi Germany during World War II in all branches of the German military. The Germans believed their Enigma system was completely secure—its often changing codes were unbreakable—or so they thought. Even many British code breakers believed the Enigma code could not be broken.

On May 9, 1941, however, crew members of the British destroyer Bulldog boarded a sinking German U-boat and recovered her cryptographic material, an Enigma machine, including cipher tables and current Enigma keys. The captured material allowed all U-boat traffic to be read for several weeks until the Germans routinely shifted the encoding keys. Though this was useful for only several weeks, having an Enigma machine in their possession allowed British cryptographers to understand the system and figure out ways to continue to break the German codes. Practically, this allowed Allied protection strategists to warn convoys where German wolf packs were lurking and then diverting them to safer waters. (Ironically, the Germans also managed to break the British naval code for a while and thus could counter defensive actions of the Allies.) Though being able to intercept German tactical radio messages some of the time did not win the Battle of the Atlantic, it was another means which helped turn the tide of the battle in favor of the Allies.

The standard weapon used to attack a submerged U-boat, particularly by the US Navy, was the conventional depth charge. They were barrel-like explosive devices, weighing about 600 pounds with a fuse set to detonate at a pre-determined depth. When detonated, a depth charge subjected a nearby submarine to a powerful and destructive hydraulic shock wave which could collapse the hull and sink the vessel. Allied destroyers, corvettes, and other escort vessels would typically carry two racks of depth charges at the stern of the vessel. When sonar detected a submarine beneath, the commanding officer would order depth charges dropped. They would roll off their rack into the water and then detonate at the prescribed depth. The depth charge did not have to make direct contact with the submarine. Other systems "shot" a depth charge off the stern of a ship and to its sides. But the principle was the same. Detonation in the vicinity of the sub would often cause sufficient damage either causing the sub to sink or make an emergency forced surfacing. Once on the surface, the heavy guns of the destroyer could finish the sub off.

However, one of the limitations of the early British ASDIC—and American sonar was that the submarine would disappear from the screen of the device when it was at close range. The British solution to this problem was the development of a system called the Hedgehog. It was a system mounted near the bow of the destroyer or corvette. It contained 24 forward-firing mortars shot out ahead of the ship and hopefully onto the nearby submerged submarine. Each projectile had a warhead of about 65 pounds. Whereas the depth charges dropped from the stern of a destroyer relied on a powerful hydraulic shock wave to damage an enemy submarine, the Hedgehog mortars did not detonate unless they made physical contact. If nothing was hit, there were no explosions. However, when explosions were heard after firing the hedgehogs, the crew of the destroyer knew the sub had been hit. A direct hit by one or two Hedgehog shells was usually sufficient to sink a German U-boat.

The Hedgehog system was deployed on British ships in mid-1942. The American Navy also adapted the system on many of its destroyers. Initially, the Hedgehogs did not prove very effective. However, when crews became properly trained, it was an effective anti-submarine weapon. By the end of the war, statistics showed that on average, one in every five attacks made by the Hedgehog system resulted in a submarine kill, compared with less than one in 80 with depth charges.

The Hedgehog system was not a decisive factor in the Battle of the Atlantic. However, along with the other technological innovations and sheer determination by sailors in both the US and Royal Navies, they together turned the tide of the battle. By the end of 1943, the U-boat menace had been largely neutralized. With the advent of the thousands of Liberty and Victory ships being produced by the United States, Hitler could not sink them fast enough to make a difference in the outcome of the war. By the end of 1943, German U-boats had become the hunted rather than the hunters. U-boats were being hunted down and sunk at numbers the Kriegsmarine could not sustain.

In all, Hitler built 1,162 U-boats, of which 785 were destroyed and the remainder surrendered, or were scuttled to avoid surrender when Germany capitulated. Of the 785 which were destroyed, 632 U-boats were sunk at sea. Truly, the hunters had become the hunted. After mid-1942, it was all downhill for the U-boat threat to the North Atlantic. There was no one decisive battle or one major hero in the Battle of the Atlantic. But the simple fact is that the Allies prevailed and the Germans failed.

No one technology won the Battle of the Atlantic. But cumulatively, they were able to neutralize Hitler's attempt to starve England into submission. After the war, surviving German submariners attributed their defeat more than anything to the deployment of the radar systems on Allied warships and aircraft.

By the end of 1942, the United States launched Operation Torch which was accomplished by sending a naval convoy over 3,000 miles from Norfolk, Virginia, to North Africa. German submarines did not molest it or even discover it. The Battle of the Atlantic had been largely won and would be completely won in the coming year. Thousands of Allied ships were sunk and tens of thousands of merchant seamen perished. But the way was now paved for the next phase of the war—sending aircraft to Britain for the upcoming European air war.

Until the Atlantic could be relatively safe for shipping, the necessary combat and support-troop buildup with the massive supplies necessary for the invasion of the continent could not occur. Until the Battle of the Atlantic had been won, there could have been no D-Day in 1944.

CHAPTER 22

Operation Torch

On November 8, 1942, the United States launched amphibious invasions of North Africa. It was the first taste of ground conflict for American troops in the European-African Theater of war. (The US Marines were already engaged in heavy combat at Guadalcanal in the South Pacific.) However, compared to the battles in Russia and later Normandy, Operation Torch would prove to be a sideshow. But important lessons would be learned and there was some strategic value in the greater scope for the European war.

Background

Benito Mussolini was the fascist dictator of Italy and would become one of very few true allies of Adolf Hitler. He, like Hitler, had grandiose dreams of invasion, conquest, and an expanded empire. He fashioned himself as a 20th century Caesar. As Hitler yearned to exalt Germany to high estate in Europe, Mussolini had similar visions of expanding an Italian empire in northern Africa.

Italy had invaded and made Libya a colony in 1911. Mussolini therefore already had significant numbers of troops there. On

October 3, 1935, Mussolini invaded Abyssinia (now Ethiopia), the African country situated on the horn of Africa. His aim was to boost Italian national prestige, damaged by Ethiopia's defeat of Italian forces in 1896. The Italian dictator saw invading Abyssinia as an opportunity to provide land for unemployed Italians and also gain more resources to deal with the Great Depression. The League of Nations did nothing.

Mussolini watched the war between England and France against Germany unfold for nine months. After observing Hitler roll over France and then cornering the British at Dunkirk, Mussolini declared war on England on June 10, 1940. He was interested in joining the winning side. In witnessing Hitler's victories, Mussolini saw an opportunity to secure additional territory for Italy.

Mussolini therefore ordered troops already in North Africa to invade Egypt. The Italian strategy was to advance from Libya along the Egyptian coast to seize the Suez Canal. The British had had military forces in Egypt since 1882. By the terms of the Anglo-Egyptian Treaty of 1936, Britain was allowed to maintain forces near and along the Suez Canal and the Red Sea. For the Brits, the Suez Canal was a vital artery for shipping supplies to and from England's eastern dominions—from India to the Far East. Keeping the Suez under their control was something they were absolutely prepared to defend.

In 1939, Italy began to threaten Egypt. Italian forces in fact attacked British forces from Libya on September 10, 1940, advancing 65 miles into Egypt. On December 8, 1940 England counterattacked and overwhelmingly defeated numerically superior Italian and Libyan forces. The Brits lost 1,900 men killed and wounded during the battle, but took 133,298 Italian and Libyan prisoners.

Meanwhile, Hitler was observing all of this from Germany. He had entered into a treaty with Italy in September of that year, known as the Tripartite Pact. Among other things, the treaty agreed that each party (Germany, Italy, and Japan) would provide for mutual assistance should any of the signatories suffer attack. Hitler's prestige was now on the line. He had publicly committed to helping Mussolini as his closest ally.

Therefore in January of 1941, Hitler established the Afrika Korps. Its express purpose was to help the Italians hang onto its colonies in North Africa after their ignominious defeat by the British. German General Erwin Rommel was thus dispatched to Libya along with the new Afrika Korps to try and salvage the deteriorating situation. Rommel was given two German divisions along with surviving Italian troops. (The Italian forces, though substantial in number were poorly led, poorly equipped, and poorly trained.)

Though Hitler was preoccupied with plans for his upcoming invasion of Russia, he gave Rommel permission to attack British positions in Egypt. Rommel's forces were stopped dead in their tracks by the British and then forced to retreat. The north-African campaign see-sawed back and forth for more than a year and a half. The concern of the British was the threat Rommel posed to the Suez Canal, and all of the Middle East if he were able to succeed. Finally, on October 23, 1942, British General Bernard Montgomery launched a massive infantry attack at El Alamein in northwestern Egypt. Rommel was forced to withdraw to the west into Libya. The stage was thus set for Operation Torch.

In July of that year (1942), American and British high level commanders in London began planning what was to be called Operation Torch. American commanders, including Chief of Staff, General George C. Marshall and Chief of Naval Operations,

Admiral Ernest King, opposed the plan. Marshall rather wanted to begin planning for an invasion of northwestern Europe, likely in France. The British opposed this believing, correctly in hind sight, that the Allies were not ready for an all-out assault against Nazi Europe. The impasse therefore was kicked up to President Roosevelt in Washington. Roosevelt gave a direct order that Operation Torch was to have priority over other operations and was to take place at the earliest possible date. This was one of only several direct orders Roosevelt gave to military commanders during the war.

The greater objective was to defeat Rommel and his Afrika Korps and eliminate the threat to the Suez Canal. Once all forces were fully landed, the British would continue to press Rommel from Egypt out of the east while American forces would land in western North Africa and drive eastward. At some point, Rommel would be trapped and defeated. Though it did not happen as easily as it seemed on paper, eventually that is what took place.

As Torch would develop, the plan would be the most extensive amphibious assault in history to that point. Three naval task forces were assembled. The western force was based in the United States and sailed from Norfolk, Virginia, on hundreds of ships. The western task force contained 35,000 green American soldiers under the command of Major General George S. Patton. Their objective was landing at Casablanca in the northwestern-Africa French colony of Morocco.

The center task force sailed from ports in the United Kingdom whose objective was Oran, Algeria, in French North Africa. This force contained 33,000 American troops under the command of Major General Lloyd Fredendall.

The Eastern Task Force also sailed from ports in Britain and whose objective was Algiers, Algeria. This force contained 39,000 both British and American troops commanded by British Lieutenant General Kenneth Anderson along with American Major General Charles W. Ryder, commander of an American infantry division.

What is remarkable is that these three task forces arrived at the same time on November 8, across thousands of miles of ocean, and coming from three separate ports of embarkation. Altogether, Operation Torch would land over 107,000 American and British troops in northwestern Africa. Three-hundred-fifty American and British warships provided escort for over 500 troop transport vessels. The overall commander of the operation was Dwight Eisenhower, his first combat command as a general officer. He set up his headquarters at Gibraltar. Almost miraculously, the Allied armadas were not molested by German U-boats. However as noted before, by the end of 1942, the U-boat menace had been considerably diminished. Moreover, the British made efforts to detract the German navy from the Torch operation by enticing their wolf packs to pursue a large convoy headed to England. German intelligence completely missed this massive operation.

Both Morocco and Algeria were Vichy French colonies. After Hitler had defeated France, he allowed a rump French state known as Vichy France—for its capital in Vichy, France. It was a semi-autonomous state, though ultimately under the control of Nazi Germany. When France fell to Hitler, the French navy, which was still formidable, had fled to North African ports and sequestered itself in the harbor at Oran and other ports. But the French had no love for the British because of what happened at Battle of Mersel-Kébir on July 3, 1940.

Then, the Royal Navy feared the French fleet would fall into German hands in the aftermath of the Allied defeat at the Battle of France. Therefore, the British planned to either neutralize or destroy the French fleet. The British Mediterranean fleet had blockaded the harbor at Oran and sent emissaries to the French Admiral, François Darlan, requesting him to surrender to the British and join forces with the Allies. Darlan assured the British that the French fleet would remain under French control. Notwithstanding his assurances, Churchill judged that the risk was too great. Therefore, when the French refused to surrender to the British, the Royal Navy bombarded the French naval base at Mersel-Kébir, adjacent to Oran. About 1,300 French sailors and other servicemen were killed in the attack. A major French battleship was sunk, along with five French destroyers.

The French thought they were acting honorably towards their former ally in terms of their armistices with Germany and Italy. The British attack was almost universally condemned in France and resentment festered for years over what was considered a betrayal by their former ally. This would all have a direct impact on how American and British forces were received when they came ashore in North Africa on November 8, 1942.

Therefore, every effort was made to make the invasion of Operation Torch appear to be completely an American operation. There was even talk of British troops wearing American uniforms, though that apparently did not happen. The Americans and British hoped the French forces in Morocco and Algeria would surrender to them and join the Allied cause. That was not to happen, at least at the first.

The Landings at Casablanca

The Western Task Force under George Patton landed before dawn on November 8, 1942, in the vicinity of Casablanca. Because Patton hoped the French would not resist, there was no preliminary naval bombardment. This turned out to be an error as French defenses caused American casualties. Meanwhile, on the night of November 7th, a pro-Allied French General attempted a coup d'état against the French command in Morocco, hoping they would surrender to the Allies the next day. His forces surrounded the villa of the French high commissioner, loyal to Vichy. However, the commissioner telephoned loyal forces, who stopped the coup. In addition, the coup attempt alerted the French to the impending Allied invasion. He immediately strengthened the French coastal defenses.

Battles raged between French defenders and arriving American troops. With the assistance of air support from American escort carriers, American troops pushed ahead, and their objectives were captured. Patton landed at 8:00 a.m. from his command ship, the heavy cruiser USS Augusta. The beachheads were secured later in the day. United States troops thence surrounded the port of Casablanca by November 10th and the city surrendered an hour before the final assault was due to take place.

The remnants of the French navy came out and attempted to intercept the American landing at Casablanca. However, aircraft from American escort carriers, along with fire from the 16 inch guns of the American battleship USS Massachusetts neutralized the French force. The French lost a cruiser, six destroyers, and six submarines in the battle. Two US destroyers were damaged.

The Landings at Oran

The Center Task force landed at several beaches near Oran, Algeria on November 8th. One of the landing groups encountered sandbars or shallow water. Once again, remnants of the French fleet escaped the harbor and attempted to attack the American task force. They were promptly sunk or driven ashore by the US Navy. Fighting continued for a day, but on November 9, Oran surrendered. Of interest is that airborne troops of the United States flew all the way from Cornwall, England, over Spain, to drop near Oran. They thereupon captured French airfields in the area. This was the first major airborne assault ever carried out by the United States. The operation did not go smoothly with some American aircraft losing their bearings over such a long distance. Other simply landed near Oran rather than dropping their paratroopers. But they accomplished their objectives.

The Landings at Algiers

Also on November 8th, British and American troops landed near Algiers. There was little French resistance. However, while attempting to land American ranger units onto the docks at Algiers to prevent the French from destroying their port facilities or scuttle their ships, there erupted stiff fighting. But 250 American Army Rangers were able to land on the docks and quickly push inland. The port of Algiers surrendered at 6 p.m. that evening.

Aftermath

Operation Torch was a complete victory for the Allies, the Americans in particular. The United States suffered 526 soldiers killed in action, the British 547 dead. But the United States had finally gone into action against Hitler, apart from sinking his submarines. The Allies established a solid beachhead in northwestern

Africa in preparation for coming action against Rommel and his Afrika Korps. Operationally, the campaign was an unqualified success in sailing 850 ships across thousands of miles of ocean without a single loss to German U-boats, 500 of those ships sailing all the way from the United States. Even more remarkable was their ability to launch coordinated attacks all on the same day—November 8, 1942—on three objectives, Casablanca, Oran, and Algiers, spread hundreds of miles apart.

One other consequence of Operation Torch was that Hitler ordered his troops in France to move into southern France, and occupy it along with the rest of France. Thus ended the rump state of Vichy, France.

North African Campaign

From November 1942 into February 1943, American and British forces moved eastward across North Africa into Tunisia. Meanwhile, the British Eighth army under Montgomery pushed Rommel from Libya into eastern Tunisia. A climactic battle would take place between US forces and Rommel at a place called Kasserine Pass in the mountains of Tunisia. As a prelude to the main battle, the Germans had laid a trap for the inexperienced American troops. The Germans feigned a retreat before the American First Armored Division. In drawing them into a mountain pass, they then opened up on the unsuspecting Americans causing heavy casualties.

However, the main battle would take place at Kasserine Pass, a two-mile-wide gap in Tunisia's Dorsal Mountains, defended by American troops. On February 20, 1943 Rommel broke through the American line inflicting devastating casualties on US forces. The Americans withdrew from their position, leaving behind most of their equipment. More than 1,000 American soldiers were killed by Rommel's attack and hundreds more were taken prisoner. Though numerically inferior, Rommel had boxed the ears and given a bloody nose to the green US Army. It was a humiliating experience. The American commander on the scene was General Floyd Fredendall.

As a consequence of the debacle, theater commander Dwight Eisenhower relieved Fredendall of his command and sent him back to the United States. In his place, General George Patton was appointed commander of the US II Corp. He immediately began to clean house, relieving other incompetent junior officers, and demanding strict military discipline of his new command. Meanwhile, Rommel was sent back to Germany to recover from deteriorating health problems.

The Allies continued to close the vise with Montgomery pushing from the east and Patton closing from the west. By now the Germans were outflanked, out-manned, and outgunned. German and Italian forces surrendered on May 13, 1943 giving up over 275,000 prisoners of war. This vast loss of veteran troops greatly reduced the military capacity of Axis forces in the region.

The psychological effect on the Allies was great after the dark years of 1941-42. Though a backwater of the war compared to what was going on in Russia, it was a stunning strategic setback for Germany. For the first time in the war, the Allies had decisively defeated the Axis, and especially the Germans, on the ground. Hitler's pride and arrogance once again overcame sound military judgment. Moreover, the stage was set for the next phase of the Mediterranean war, the invasion of Sicily and then Italy itself.

CHAPTER 23

The Air war Over Europe

The next major Allied campaign of World War II which had to be won was the air war. Military aviation was still in its youth when World War II broke out. Strategic theories were floated and tried: some worked, some did not. For the first time since the inception of military aviation just several decades earlier, strategic bombers were being used. Fighter-interceptors would also come to maturity during World War II.

As first the RAF and then as the United States Army Air Force (USAAF) entered World War II, there were theories of air doctrine which influenced the building of aircraft and shaped strategic thinking. Some prominent air strategists in the 1930s proclaimed the next war could be won by strategic bombing alone without the need for ground forces. An Italian air-war theorist, Giulio Douhet, was widely quoted to the affect: "the bombers will always get through." It did not take long for that theory to be "shot down." As World War II developed, first British and then American bombers were shot down in terrible numbers. But Douhet's theory influenced American aviation thought nevertheless, particularly prior to and in the early years of the war.

A corollary of Douhet's doctrine asserted by United States air strategists was that American heavy bombers could defend themselves in battle. For example, the ubiquitous B-17 American heavy bomber carried twelve .50 caliber, heavy machine guns placed strategically around the aircraft in turrets or gun ports. The American B-24 heavy bomber carried similar armaments. American doctrine was to fly these airborne fortresses in "box" formations of 6 or 7 bombers in which—in theory—all the machine guns of the formation could be trained on an attacking enemy fighter. In reality, Germans fighters were so numerous and fast, the gunners on the big bombers could barely cope with them. They certainly shot down enemy fighters, but just as many bombers were shot down by the fighters in the process. Moreover, German anti-aircraft artillery shot down more Allied bombers than did German fighters. In 1944 alone German anti-aircraft guns accounted for 3,501 American planes destroyed, whereas German fighters shot down about 2,900 planes. The bombers did not always get through.

The British used several models of bombers, initially the Handley Page, the four-engine Stirling, and then the heavy, four-engine Lancaster bomber. British crews loved the Lancaster, calling it the "Lankie" which was introduced into battle in February of 1942.

The Bombing of Germany

Once it became clear that Germany was a threat in the 1930s, the RAF had started on a large expansion program with many airfields being set up. In 1934, the RAF had 42 squadrons with 800 aircraft. By 1939, that had expanded to 157 squadrons with 3,700 aircraft. Their medium bombers could reach the Ruhr region of Germany. With their long range, heavy, four-engine bombers soon to come online, they could penetrate all the way to Berlin.

In the spring of 1940 Britain had attempted an invasion of France, only to be utterly defeated, culminating in the evacuation of British troops at Dunkirk. Their only recourse was to begin bombing Germany. Though initially attempting to bomb by day, the British quickly abandoned that strategy and switched to bombing at night. Their bombers were less susceptible to being shot down by cover of darkness, though the Germans soon perfected effective night-time antiaircraft defenses. The Brits made no attempt to pinpoint military targets, but rather did what they called "area bombing" also known as carpet bombing. Accordingly, they devastated many civilian areas near military installations.

Night bombing by nature was therefore imprecise which resulted in carpet bombing of German cities which killed many civilians. The British defended this because (1) the Germans were doing the same to England and (2) they hoped to so damage German cities that the population would rise up and demand the war to end. That never happened.

Nevertheless, the head of Bomber Command, General Arthur Harris, believed that bombing alone could bring Germany to its knees and render a land invasion of Germany unnecessary. He therefore sought to systematically obliterate German industrial cities. He was pitiless in bombing the Third Reich and came to be known as "Bomber Harris." He once said, "The Nazis entered this war under the rather childish delusion that they were going to bomb everyone else, and nobody was going to bomb them. At Rotterdam, London, and Warsaw, they put their theory into operation. They sowed the wind and now they are going to reap the whirlwind."

Because of international criticism, even from America, British military planners were faced with this dilemma: abandon all

offensive action against Germany or bomb German cities and civilians in hope of damaging German industry and the morale of
the German people. The Brits chose the latter of "area bombing"
and pursued it with ferocity. Yet, it was a bloody business even
for the British. Tens of thousands of British airmen were shot
down over Germany.

On the night of March 28-29, 1942, two-hundred-thirty-four aircraft bombed the German port of Lübeck. This target was chosen
not because of any significant military value, but because it was
expected to be particularly flammable—in Harris's words, "built
more like a fire lighter than a city." The ancient wooden buildings of the city burned quickly, and the raid destroyed most of
the city's center.

On May 30, 1942, in the wee hours of the night 1,046 British
bombers dropped over 2,000 tons of high explosive and incendi-
aries on the medieval city of Cologne. The resulting fires burned
it to the ground, from one end to the other. The damage was
almost complete. The fires could be seen 600 miles away at an
altitude of 20,000 feet. Some 3,300 houses were destroyed, and
10,000 were damaged. Twelve thousand separate fires raged,
destroying 36 factories, damaging 270 more, and leaving 45,000
people with no place to live or to work. Only 384 civilians and 85
soldiers were killed, but thousands were forced to evacuate the
city. Bomber Command lost 40 bombers.

Hamburg, Germany, suffered a similar fate. On the evening
of July 24th, British aircraft dropped 2,300 tons of incendiary
bombs on Hamburg. The explosive power was the equivalent of
what German bombers had dropped on London in their five most
destructive raids. More than 1,500 German civilians were killed
in what was the first British raid against the city. Britain lost
only 12 bombers thanks to a new radar-jamming device called

"Window," which consisted of strips of aluminum foil dropped by the bombers en route to their target. These Window strips confused German radar, which mistook the strips for dozens and dozens of aircraft, diverting them from the flight-path of the actual bombers.

The Americans Arrive

In mid-1942, the United States Army Air Force (USAAF) began arriving in England. Their primary weapon was the Boeing B-17 Flying Fortress. Ironically, of all the combatant nations in World War I, only the British and the Americans had developed long-range, four-engine, heavy bombers. A sister aircraft was the Consolidated B-24 Liberator. It was likewise a four-engine, long-range, heavy bomber. The USAAF command in England for bombing Germany was the US Eighth Air Force, commanded initially by General Carl Spaatz and later in the war by General Jimmy Doolittle. The Eighth Air Force made several small raids into France late in 1942, but did not begin heavy bombing of Germany until well into 1943.

The Air War in Europe eventually would be decisively won by the Allies. Without that victory, D-Day could not have succeeded. But the unsung hero of the air war was the American industrial might back home. Various aircraft manufacturers turned out massive numbers of military aircraft. American military industrial production was overwhelming. Over the war years, 12,731 B-17 Flying Fortresses were built as well as 18,000 B-24 Liberators. Over 15,000 P-47 fighters bombers were produced along with 15,000 P-51s, not to mention numerous other types of military aircraft. By the end of the war, America overall produced 300,000 military aircraft of all types. No war was more industrialized than World War II. It was a war won as much by machine shops as by machine guns.

Precision Bombing?

The Americans came with a completely different philosophy and doctrine of strategic bombing. They viewed the British concept of area or carpet bombing of civilian areas as immoral. Rather, the American doctrine of bombing was "precision" bombing. The USAAF intended to fly its missions over Germany in broaD-Daylight and not at night. The plan was to only target sites of military value such as munitions factories, military installations, oil refineries, and anything which aided the German war effort. The British, who had already had two years of combat experience in the air over Germany thought the inexperienced Americans naive, stubborn, and foolish. As a result, there was not a little friction between the two command structures. The Brits had trieD-Daylight bombing and quickly gave up on it due to heavy losses.

Norden Bombsight

The Americans arrived with a top-secret, precision aiming device known as the Norden bombsight. They claimed they could drop a bomb in a pickle barrel from 30,000 feet with the Norden. Ironically, Carl Norden, the Dutch engineer and designer, had offered his bombsight to the Germans before the war. They were not interested. It therefore was not all that top secret, as things turned out.

The Norden bombsight was a sophisticated mechanical computer. The bombardier would input various flight data such as altitude, airspeed, and presumed atmospheric wind speed. As the aircraft approached the target, the Norden would take over control of the aircraft for a short period—as an autopilot, keeping the approach to the target straight and level. The bombsight would then calculate the trajectory to the target and release the bombs automatically at the precise right moment. In fact, under *ideal* conditions, the Norden Bombsight was thought to be able to place bombs within 100 feet of the target from an altitude 20,000 feet. But even an error of 100 feet was still basically a miss. In reality, most American bombers flew higher than that, incurring even greater error.

However, conditions were rarely ideal. For the Norden to work, there had to be clear skies over the target. From mid-Autumn through Spring, skies over northern Europe were often overcast with low ceilings. As the war developed, the Germans developed fierce air defenses, including world-class fighter interceptors and heavy anti-aircraft fire (flak). The force of flak explosions would literally bounce bombers around, even if they were not actually hit. German fighters forced bombers to jink and jerk to avoid being hit. But the primary design flaw in the World War II models of the Norden bombsight was little or no understanding of the jet stream. Meteorology in the 1930s did not well understand the dynamics thereof. As American bombers routinely flew at 30,000 feet to avoid anti-aircraft defenses, the strong velocity of the jet stream in addition to the air speed of the aircraft simply blew bombs off their target as they cascaded down through thousands of feet of altitude on their way to ground.

The long and the short of it was the Norden Bombsight, for all its sophistication and secrecy, could not place bombs with any precision whatsoever. This went to the very heart of the American doctrine of precision bombing. Though unintended, American bombs fell over widely scattered areas near targets, including civilian neighborhoods, and wound up being little different in practice than the British on-purpose area bombing. Some bombs hit targets, though most missed. The basic solution was to send more bombers with more bombs. Eventually enough of them began to cause the intended damage to targets of strategic or military value.

Crude target acquisition radar sets were tried, but not with much success. The bottom line is that American bombing in the first year and a half over Germany was ineffectual. Targets were hit by chance more than on purpose.

Round the Clock

And so the British bombed Germany by night and the Americans by day. The Allied bombing became literally round-the-clock bombing. On many occasions, the Americans came back and bombed the same area the Brits had the night before, trying to completely destroy the target. The only way to put a target out of action permanently was to bomb it incessantly.

In March of 1943, the RAF began massive raids against the Ruhr Valley of Germany. Most of German's heavy industry was located in this region. Night after night, British bombers hit industrial cities in the Ruhr. Regarding his bombing campaign against the Ruhr, Bomber Harris said, "It will cost us 500 aircraft. It will cost the Germans the war." It did not turn out that way, but German industry was severely damaged.

Bombing Berlin

During the winter of 1943-44, the RAF began to regularly bomb Berlin. Harris declared his goal was to strike at the enemy's capital, "to burn his black heart out." But Berlin was not like old medieval German cities such as Cologne or Hamburg. It was a relatively modern city with planned neighborhoods, broad streets, and wide boulevards. Many of its more modern buildings were constructed of masonry or concrete. Hence, there were not the firestorms which developed in other places. The wide streets became fire-breaks, helping to contain the fires started by the bombs. Nevertheless, 14,000 civilians in Berlin were killed or injured that year. To Harris, this was not a problem, but rather something the RAF should exacerbate. His plan was to do much damage to Germany and destroy German morale. However, the Berlin campaign cost the British dearly in aircraft and crews lost.

German air defenses

Germany quickly fielded an impressive and effective anti-aircraft defense system. Two principal weapons were in their arsenal. First, a powerful piece of artillery known as their 88 mm gun. Then, they also developed several very effective fighter planes, the Me 109 and the Fw 190. They certainly used other fighters, including a jet model at the end of the war, but they either were ineffective or arrived too late to make any difference in the war. Additionally, flak towers rising ten and more stories and constructed of heavy steel reinforced concrete were built particularly in and around Berlin, Hamburg, and Vienna. They bristled with various calibers of flak guns.

88 mm Anti-Aircraft artillery

Likely the most powerful and effective artillery piece used by any combatant in World War II was the German 88mm anti-aircraft gun developed in the 1930s. Though they certainly used other types of anti-aircraft artillery, the 88 mm was one of the most recognized German weapons of the war. The Germans initially called it the 8.8 cm Flak 18, though there would be later improved models. Flak is a contraction of the German word *Flugabwehrkanone* (also referred to as *Fliegerabwehrkanone*) meaning "aircraftdefense cannon." In English, "flak" became a generic term for ground-based antiaircraft fire. In casual use, the guns were universally known as the Achtacht ("eighteights") by Germans and the "eightyeights" by the Americans.

The 88s could fire explosive shells as high as 32,000 feet, though its most effective range was about 26,000 feet. An 88 shell did not have to actually hit a bomber. Rather it merely needed to explode in the vicinity with the shrapnel from the shell tearing into any man or machine nearby. Over twenty-one thousand 88 guns were produced before and during the war, manufactured with high-grade Krupp steel.

The design was so versatile that Germans quickly discovered it to be a powerful anti-tank weapon. No American tank could withstand a direct hit by an 88 shell. The Germans would later install 88mm cannons into their Tiger and Panther tanks, making them especially formidable. An American military historian years later would write, American troops "knew that the greatest single weapon of the war, the atomic bomb excepted, was the German 88 mm flattrajectory gun, which brought down thousands of bombers and tens of thousands of soldiers. The Allies had nothing as good."

German fighter/interceptors

Me 109

Nazi Germany produced numerous high-quality military aircraft during the war. For its air defense system, however, two fighters were the work horses of the Luftwaffe. The first was the Messerschmitt Me 109, produced by Willie Messerschmitt, a German aeronautical engineer. The Me 109 was designed on the principle that later defined American muscle cars: a small airframe with a very powerful engine. The Me 109 was arguably one of the finest fighter planes of World War II. Messerschmitt's 109 was built in greater numbers than any other fighter plane, with over 33,000 produced. It shot down more Allied planes than any other German aircraft. Ironically, it was also used by the infant Israeli air force in their war of independence in 1948.

The Me 109 later went up against newer and more powerful adversaries such as upgraded British Spitfires and the P-51 Mustang. But even in the hands of a skilled pilot, the Me 109 could not hold its own against the American P-51. The Me 109 nevertheless was a formidable adversary to Allied air forces.

Fw 190

The other workhorse fighter of the Luftwaffe was the Focke-Wulf Fw 190. Along with the Me 109, the Fw 190 formed the backbone of the German fighter force. The Fw 190 was the only completely successful piston-driven fighter introduced by the Luftwaffe after World War II started. Produced by the Focke-Wulf company, it was unlike the Me 109 in that it had a powerful *radial* engine which was less susceptible to damage by enemy gunfire. Over 20,000 were built during the war.

The Fw 190 had a top speed of about 410 miles per hour with a ceiling of 35,000 feet. The fighter's heavy cannon armaments made it a potent weapon against Allied bombers. It played a major role in turning back the US Army Air Force's unescorteD-Daylight bombardment offensive in the summer and autumn of 1943.

The Fw 190's effectiveness was diminished with the appearance of large quantities of drop-tank-equipped P-38 Lightnings and P-47 Thunderbolts in late 1943. The 190 could not match the performance of these turbo-supercharged US fighters above 30,000 feet. The appearance in 1944 of large numbers of P-51 Mustangs put the Fw 190 at a permanent disadvantage. Later variants of the Fw 190 were introduced in 1944, increasing its performance. But few of them were produced to make much of a difference.

The Tide Begins to Turn

Beginning slowly at the end of 1942 and gaining momentum throughout 1943, the USAAF bombing campaign began to hit its stride. US bombers ranged across Germany, but with terrible losses. The American point system was that when a bomber crew member had flown 25 combat (i.e., bombing) missions over Germany, he could return to the United States. Only a small percentage of bomber crews achieved that goal. Most were shot down long before. Between Luftwaffe fighters and ever increasingly accurate anti-aircraft artillery (flak), every bombing mission became a life and death gauntlet for the bomber crews.

On August 17, 1943, the Eighth Air Force sent 184 bombers from bases in England to bomb ball-bearing factories in Schweinfurt, Germany. Ball bearings were critical to virtually every facet of German armament, from tanks to aircraft. If the source of the ball bearings could be destroyed, the German war machine would soon grind to a halt—or so was the theory. The American bombers faced attack by German fighters all the way there and virtually all the way back to England. Loss of aircraft and crews were high. Although the inflicted damage was not as extensive as the US Eighth Air Force had wanted, it did temporarily disrupt about 34% of the bearing production capacity. Another even more notable raid on Schweinfurt took place October 14, 1943, during which an even higher rate of loss was suffered—60 of the 229 attacking B17 bombers were lost. It came to be known as "Black Thursday" and would temporarily put such attacks by the USAAF on German industrial centers on hold until long-range escort fighters became available.

Ironically, the Germans, though repairing their bombed out factories, determined to disperse ball bearing production throughout the country. Furthermore, they had adequate stock piles located throughout the nation which kept their armament industries operational. Though not known until years after the war, the "neutral" Swiss began supplying German with ball bearings, at a handsome profit, or course. So, the raids on Schweinfurt had little effect on the German war machine, but it cost the Allies heavily in downed aircraft and crews lost.

Escort Fighters

The Eighth Air Force endeavored to provide escort fighters for their bombing missions. However, none of them had the range to accompany the big bombers all the way into Germany and back. But in early 1944, the P-51 Mustang, P-38 Lightnings, and P-47 Thunderbolt with drop fuel tanks began escorting bombing missions. The tide of the air war began to change dramatically. The P-51 Mustang became the premier fighter aircraft of World War II and probably the finest piston-driven fighter aircraft ever built. With its drop fuel tanks it could go the distance into Germany and back. It was very fast, maneuverable, and was superior to anything the Luftwaffe threw against it.

The Luftwaffe's twin-engine Messerschmitt Me 110 heavy fighter, sent up to attack American bombers, proved to be easy prey for the Mustangs. The 110s were quickly withdrawn from combat. The FockeWulf Fw 190A, with poor high-altitude performance, was outperformed by the Mustang and suffered heavy losses. Ironically, the P-51 had been around since 1942, but its original American Allison engine was underwhelming in performance. The RAF rather replaced the Allison engines with British Roll-Royce Merlin engines. Overnight, the P-51 went from being a mediocre fighter to the best in the world. It was instrumental in turning the tide of the battle.

The 15th Air Force

Meanwhile, American and British forces had invaded Italy. As they slogged their way up the Italian peninsula, they captured the German-Italian Foggia Airfield Complex near Foggia in eastern Italy. Immediately, the USAF 15th Air Force moved in and began operations. While the Eighth Air Force attacked German targets from England to the northwest, the 15th Air Force began to attack critical targets from Italy to the south. The 15th would fly primarily B-24 Liberator bombers.

Germany had few petroleum resources, but had to import almost all of its oil. One major source for the Nazi war machine was the oil fields in the vicinity of the Ploiesti, Romania. The oil fields were attacked on April 5, 1944 and again on April 15th and 24th, inflicting additional damage. The Americans suffered terrible losses in the attacks, but the Ploiesti fields were heavily damaged. The Germans tried to repair the damage only to be bombed again. The US high command realized that Germany's war machine ran on oil (i.e., gasoline, diesel, fuel oil, etc.). Therefore, American strategic bombing turned its focus to destroying the German oil industry. Attacks on oil targets assumed top priority. By October of 1944, vast fleets of heavy bombers, escorted by P-38 Lightning and P-51 Mustang fighters, attacked refineries in Germany, Czechoslovakia, and Romania. The P-51 escorts were able to establish an environment of air superiority, enabling the bombers to roam widely across southern and Eastern Europe.

Running out of Gas

German scientists had devised a way to produce petroleum products from coal which Germany had in abundance. The synthetic fuels were inferior to natural fuels, but sufficed. However, such synthetic fuels were much more expensive and were not produced in large quantities. American bomber fleets thus began to destroy the German synthetic oil facilities. Little by little, Germany was running out of gas, literally. By the end of the war, German production of gasoline and other fuels was down to 5% of its pre-war levels. As 1944 turned into 1945, this had a profound impact on the Nazi war machine. Aviation fuel for fighters became so scarce that new pilots received only a fraction of the in-air training compared to American and British pilots. Their lack of training made them sitting ducks for the experienced, battle-hardened American fighter pilots. German tanks began to run out of fuel. In the Battle of the Bulge in the winter of 1944-45, German tanks had only enough fuel to run for several days or so. Their desperate plan was to capture American fuel supplies or siphon fuel from captured American vehicles.

As the Luftwaffe's fuel supplies dried up, it was reduced to assigning its flight crews to anti-aircraft batteries or sending them to infantry units. In 1944 the Luftwaffe operated 39,000 flak batteries staffed by up to one million men and women. Additionally, by 1944, a large proportion of Germany's war expenditures was channeled to air defense. This was particularly critical as the resurgent Red Army in Russia was driving westward and as the Allies landed at Normandy.

The End of the Luftwaffe

Early in 1944, General James Doolittle, the new commander of the Eighth Air Force, ordered many American fighter pilots to stop flying in formation with the bombers and instead attack the Luftwaffe wherever it could be found. Mustang groups were sent far ahead of the bombers in "fighter sweeps" in order to intercept attacking German fighters. In late February 1944, Eighth Air Force fighter units began systematic strafing attacks on German airfields with increasing frequency and intensity through the Spring. Their objective was to achieve air supremacy before the upcoming Normandy battlefield. These attacks were usually by P-51s returning from escorting bombing missions. Doolittle ordered them to "go hunting for Jerries. Flush them out in the air and beat them up on the ground on the way home."

In March, many groups of P-51s were assigned to directly attack Luftwaffe airfields. As a result, large numbers of Luftwaffe aircraft were destroyed on the ground. They did not have fuel to come up and fight. The few that did were summarily shot down. By the time of Operation Overlord (D-Day) in June, the Luftwaffe had basically been eliminated from northwestern Europe. In fact on D-Day itself, only two German fighters appeared over the beaches, fired a few rounds, and headed for safety. For all practical purpose, the Luftwaffe ceased to exist, at least in France and Belgium by the summer of 1944. One bitter joke amongst German ground troops was that British aircraft were blue, American planes were silver, but the Luftwaffe was invisible.

On April 15, 1944, Eighth Air Force Fighter Command began "Operation Jackpot," with attacks on Luftwaffe fighter airfields. As the success of these missions increased, the number of fighters at the German airbases fell to the point where they were no longer considered worthwhile targets.

Close Ground Support

As the time of D-Day approached (June 6, 1944), Eisenhower, over the strenuous opposition of both British and American air commanders, ordered strategic bombing over Germany halted. Ike directed that in preparation for Operation Overlord, Allied heavy bombers were to begin softening up German defenses in northwestern France. Anything of military value was to be bombed from railroad marshaling yards, to bridges, and of course military installations.

On May 1, 1944, over 1,300 Eighth Air Force heavy bombers made all-out attacks on the French rail network, striking targets in France and Belgium. On May 7th, another 1,000 bombers hit additional targets along the French coast, striking fortifications, bridges and marshaling areas. On D-Day, over 2,300 sorties were flown by Eighth Air Force heavy bombers in the Normandy and Cherbourg invasion areas, all aimed at neutralizing enemy coastal defenses and frontline troops. Once again, precision bombing or the lack thereof was a problem. Though thousands of tons of bombs were dropped, near the Normandy beaches, not many hit their targets. German installations in the area were discomfitted, but not many were eliminated.

After the Normandy invasion, Allied troops quickly became bogged down. The American sectors beyond the beaches were a maze of heavy, thick, ancient hedgerows called the bocage. British forces to the east were stymied by German opposition and the sluggish leadership of General Montgomery. Finally, the American commander on the ground, General Omar Bradley ordered Operation Cobra. It commenced on July 25, 1944, with Bradley directing massive heavy bomber attacks of the German positions holding back an American breakout of Normandy. He targeted an area 6,000 yards long by 2,200 yards wide (or, 3.4

miles by 1.3 miles) of the German positions which were stymy-ing the Americans. That day, 1,500 heavy bombers, 380 medium bombers, and 550 fighter bombers dropped 4,000 tons of high explosives on the concentrated German forces.

However, once again precision bombing or the lack thereof be-came a tragic issue. Seventy-seven American bombers dropped their bombs short of the German lines, falling upon American troops. Sadly, 111 American troops were killed and 490 were wounded. The dead included Bradley's friend General Lesley McNair—the highest-ranking US soldier to be killed in action in the European Theater of Operations.

The Germans were stunned senseless, with tanks overturned, telephone wires severed, commanders missing, and a third of their combat troops killed or wounded. The German defensive line was broken and American commanders rushed their troops forward. The Germans retreated in a rout. The American break-out of Normandy had begun. Because of the lack of precision bombing, American ground commanders would never again ask for strategic bombers for close ground support. The Eighth Air Force thereafter resumed its bombing of Germany.

The Ninth Air Force

In the early Spring of 1944, the United State Army Air Force was reorganized as the United States Strategic Air Forces (USSTAF). The practical outworking of this was tactical fighter bombers squadrons were thence organized as the US Ninth Air Force. The Eighth Air Force would continue its strategic bombing of Germany, whereas the Ninth Air Force would be focused on close air support of the coming Normandy invasion.

In the summer 1944, the US Ninth Air Force started operating out of newly acquired bases in France. Though some P-51s and P-38s were involved, the workhorse of the Ninth Air Force was the P-47 Thunderbolt. It was called the "Jug" by the airmen because their homely appearance resembled old fashion milk jugs. As the American Army charged forward into France, the Ninth Air Force flew ahead of them, shooting up or bombing anything on the ground that moved—or could move. In the final year of the war, they destroyed 86,000 railroad cars, 9,000 locomotives, 68,000 trucks, and 6,000 German tanks and artillery pieces. The P-47 Thunderbolts alone dropped 120,000 tons of bombs and thousands of canisters of napalm. They fired 135 million .50 caliber heavy machine gun bullets along with 60,000 rockets. They also shot up 4,000 enemy planes.

Beyond the destruction imposed, the flights of unopposed Allied fighter-bombers ruined morale of the Germans, as privates and generals alike dived into ditches for cover. Field Marshal Erwin Rommel, himself, was critically wounded, almost fatally, when he rode across France in his command car in broaD-Daylight.

One German commander would later lament that the Americans have "complete mastery of the air. They bomb and strafe every movement, even single vehicles and individuals. They reconnoiter our area constantly and direct their artillery fire . . .. The feeling of helplessness against enemy aircraft has a paralyzing effect, and during the bombing barrage the effect on inexperienced troops is literally 'soulshattering.'"

The American Ninth Air Force was a major factor in the Allied victories across France in 1944. Without it, neither Patton's armies nor other Allied commanders would have achieved the victories which they did.

Meanwhile, the American 8th and 15th strategic air forces as well as the British Bomber Command continued their bombing of Germany. As 1944 became 1945, German oil resources were virtually destroyed. Any city of any size, particularly in western Germany had been reduced to rubble. The Luftwaffe was eradicated. All that was left was for Allied armies to cross the Rhine River into Germany proper, along with the Russians from the east, and finish the war. That they did in April and early May of 1945.

Assessment

Historians ever since have debated the effectiveness of the strategic bombing of Germany. Most of the theories of air warfare and bombing prior to the war were found lacking. Air power did not force Germany out of the war. Precision bombing was never achieved to any great degree. Even heavily armored American bombers could not defend themselves by themselves. The first several years of the air war was as much symbolic as it was effective. The Allies were attacking Germany, though the initial results were marginal. However, as the bombing continued, German resources little by little began to be destroyed. As more and more American and British bombers entered the fray, Germany could not replenish or repair the damage done. The coup de grâce was the destruction of German oil resources by strategic bombing. As a result, Hitler's war machine literally ran out of gas. Moreover, the fierceness of low-level ground attack by the Ninth Air Force from mid-1944 onward was a significant factor in the victories of Allied ground forces in the latter part of 1944.

One quarter of Hitler's war economy was destroyed because of direct bomb damage, resulting in delays, shortages, and round-about solutions. This necessitated increased spending on anti-aircraft systems, civil defense, repair, and moving factories to safer locations. The raids were so extensive and oft repeated, that in city after city the repair systems broke down. The bombing stopped the full mobilization of the German economy. Nazi Minister of Armaments Albert Speer and his staff were efficient in devising solutions and workarounds, but their problems became more difficult every week as one backup system after another failed. By March 1945, most of Germany's factories, railroads, and telephones had stopped working. Troops, tanks, trains, and trucks could not move. About 25,000 civilians died in the Dresden air raid on February 13-14, where another firestorm incinerated the city.

After the war Albert Speer, Hitler's Minister of Armaments remarked what broke the back of Germany was not so much the Allied bombing of German industries. Rather, it was the bombing of the Nazi's oil supply system—refineries, synthetic oil plants, and the infrastructure to transport oil products. As noted above, the Allied bombing thereof caused Germany to literally run out of gas, in the final months of the war.

However, the ultimately victorious air war for the Allies was costly. Close to 80,000 American airmen perished in battle over Europe. The English lost a similar number. Airmen died at a higher rate per capita than did infantry soldiers. The United States lost almost 10,000 bombers and over 8,000 fighters. Various numbers are given estimating civilians killed by Allied bombing in World War II, but they range from 400,000 to 800,000 deaths.

All in all, World War II would have ended far differently had it not been for the Allied strategic bombing of Germany. The same can be said for the tactical operations of the American Ninth Air Force. The second major phase of the American war against Germany had been won. For all his bravado and arrogance, Hitler had bitten off more than he could chew when he blundered by declaring war on the United States. The way was now paved for the third phase—the ground invasion of Europe in June 1944.

CHAPTER 24

The Ground War in Europe

By the end of 1943, the Allies had, for the most part, won the Battle of the Atlantic. The sea lanes, though still perilous, witnessed a dramatic shift. Allied cooperation and coordination had driven Germany's U-boats from being the hunters to the hunted. Moreover, the air war over Europe was quickly resolving in favor of the Allies. The Luftwaffe by 1944 had been reduced to basically the defense of Germany and by later that year, would no longer be a force to be reckoned with even there. The next phase of the war, at least from the American perspective, was the coming land battles in Europe. The invasion of France and the European continent by Allied forces loomed ahead. However, before the main invasion, there would be several prelude actions, including the invasion of Sicily and Italy.

After the defeat of the Afrika Korps in North Africa at the end 1942 and early 43, the Allies began considering their next steps. At the Casablanca Conference in January 1943, top British and American leadership began to plan their strategy for the coming year. The British favored invading Sicily. The Americans were cool to the idea. From their perspective, the United States thought taking Sicily to be irrelevant to the greater objective of defeating Nazi Germany. The British however, argued that taking Sicily and Italy would force Germany to further extend its forces, particularly when they were in serious trouble in Russia. Also, Italy could be knocked out of the war and Turkey might even join the Allies. The Americans were finally persuaded on the grounds that greater protection would result for Allied shipping through the Mediterranean by the removal of Axis air and naval forces from Sicily and Italy.

Sicily

The planned invasion of Sicily therefore began on July 9, 1943. Its code name was Operation Husky. British and American land, air, and sea forces coordinated for amphibious landings on the southeast corner of the island. It was a huge operation with 160,000 Allied personnel involved, including over 3,000 ships and 4,000 aircraft.

The invading Allies landed against a poorly prepared enemy. Allied deception had confused the Axis leadership into thinking their next move would be against Greece, or maybe Sardinia. Furthermore, distrust and dislike between the Nazis and the Italians hampered their cooperation and coordination. Mussolini insisted that the Axis forces on Sicily remain under the control of the Italian general, Alfredo Guzzoni.

Overall commander of the Allied operation was Dwight Eisenhower. The American Seventh Army under George Patton along with the British Eighth army under Bernard Montgomery landed on separate beaches on the southeast corner of the island. The plan was for them to support one another against an Axis counterattack.

On invasion day, July 9, 1943, the weather was more formidable than the Italians or Germans. A fierce summer storm almost prompted commanders to abort the operation. Sea conditions made many soldiers severely seasick, only adding to their troubles. The weather caused further difficulties for even the experienced British First Airborne and the American 82nd Airborne divisions. They were widely scattered by the storm and many airborne gliders wound up coming down in the sea, drowning all on board. The only good thing about the weather is that the Axis defenders on the island did not think landings were possible in such conditions, thus letting their guard down.

That morning, Allied forces came ashore to relatively light resistance, with only light casualties recorded. The British objective was to secure the port of Syracuse which they accomplished with hardly a fight. Montgomery was then ordered to advance up the east coast of Sicily with Patton covering his flank to the west, a role against which Patton chafed. Montgomery however, soon bogged down as the Germans brought up additional troops while relegating the unenthusiastic Italian forces to the rear.

Patton rather seized the initiative and drove northwestward toward Palermo on the north coast of the Island. Interestingly, he was assisted by local Mafiosi who detested Mussolini and viewed American soldiers as liberators. German forces on the island retreated eastward to the port city of Messina. Meanwhile, Montgomery slowly plodded northward along the east coast of the island, while Patton charged in hot pursuit after the retreating Germans along the north shore of Sicily. Through failure of the Allied high command, the Germans were able to evacuate 40,000 troops and about 60,000 Italian troops across the straits of Messina to the Italian mainland. There was no attempt by the Allied navies or air forces to interdict their escape. Those forces would come back to haunt Allied forces when they landed in

Italy proper. Their escape contributed greatly to the hard battles the Allies would soon face on the Italian mainland.

Though Montgomery had a far shorter distance to march to Messina than Patton, Patton beat him, with unabashed one-up-manship. Patton took deep pleasure in seeing his troops enter Messina 17 hours before Montgomery arrived.

By the end of the battle, the Allies had landed 467,000 personnel on Sicily, suffering over 5,500 killed in action and another 17,000 wounded or missing in action. The Axis forces suffered almost twice those numbers.

Notwithstanding the escape of Axis forces from Sicily, the campaign was otherwise quite successful. Mussolini was already increasingly unpopular amongst his own people. Food shortages, Allied air raids on Italian cities, and the unpopular presence of German troops all combined for unrest across the nation. The invasion of Sicily was the final blow to Mussolini's fading prestige. On July 24, 1943, Italian King Victor Emmanuel deposed Mussolini (a.k.a, the Duce). He was arrested and detained while a new head of state, Marshal Pietro Badoglio, was appointed by the king. Though outwardly professing loyalty to Germany, the Italian government made private peace overtures to the Allies. They signed a separate armistice on September 3, 1943. Whereupon, the Germans shifted from being allies of Italy to occupiers. They vented their wrath against the Italian populace and became an enemy of their former allies.

The Invasion of Italy

With victory in North Africa and Sicily, the Allies decided to invade Italy anticipating relatively easy battles. However, the Italian campaign would prove to be excruciating in its difficulty.

British forces under Montgomery invaded on September 3, 1943, crossing the Strait of Messina to Calabria on the toe of the Italian boot without opposition. The British zone of operation in Italy would primarily be moving up the east coast of the peninsula, while the Americans would attempt to move up the west coast. Six days later, American forces landed at Salerno 150 miles to the north of the British landing. The plan was straight forward. The Americans near Salerno would establish a line across the Italian peninsula while the British army chased the Germans toward it from the south. Allied commanders anticipated relatively easy battles.

However, the Salerno operation went badly. On September 9th, the US Fifth Army, under General Mark W. Clark, expecting weak resistance, landed in Operation Avalanche and came up against heavy resistance. Clark's forces found themselves up against the veteran, battletested, 16th Panzer Division. They were dug into strong points along the beaches at Salerno, with artillery emplaced on the high ground.

Clark's force was soon in serious trouble. He had only three divisions on land with one other on the way. The Wehrmacht moved quickly, rushing no fewer than six divisions to the front in the first two days. German defenses were so skillfully prepared, that after four days the Americans were still trapped in a shallow beachhead. German Panzer units came to within one mile of the Allied beachhead. But, allied artillery, naval bombardment from US Navy warships lying just offshore, and bombing by B17s all poured a rain of steel down on the Germans, preventing them from crushing Clark's beachhead. It was not a valiant victory, but Clark's force survived. It was a precursor of the battles ahead.

Coinciding with the British landings, the new Italian government, led by Pietro Badoglio, signed a secret agreement on September

3rd promising that Italian forces would stand down, leaving only German resistance. It was publicly announced one day before the Allied landing at Salerno on September 8th. Upon learning of Italy's unilateral armistice with the Allies, the Germans promptly seized Italian military installations, imprisoned their hapless former allies, and fired on unsuspecting Italian ships. By the time the Americans established themselves ashore, the Germans had moved north and had established new defensive lines. This became known as the Gustav Line, about 75 miles north of Naples.

Several factors were now at work, strengthening German positions. As the Allies advanced, they encountered increasingly difficult terrain, particularly the Apennine Mountains forming a rugged spine down the middle of the Italian peninsula. The difficult terrain of the Italian peninsula made progress for the invading armies slow. During the campaign, British combat engineers were forced to build 3,618 bridges across fast moving rivers and rugged terrain. Moreover, German reinforcements were arriving just as the Allies began diverting forces to England in preparation for the upcoming crosschannel invasion of France.

The Germans retreated slowly, building a series of skillfully prepared defensive lines across the peninsula. As each line was painfully breached by the Allies, the Germans fell back to their next line.

The Allies therefore attempted to break the deadlock in January 1944 by doing an end-around play. Under General Lucas, the Americans landed new forces at Anzio, 50 miles north of the German Gustav line and 30 miles south of Rome. The attack should have been successful. The Anzio landing was preceded by a diversionary attack against the Gustav Line at Cassino. German troops stationed at Rome were therefore sent to strengthen the Gustav line, leaving Anzio landing virtually undefended.

The American's plan was that once the landing was successfully made, the Germans would be forced to send divisions from the Gustav line back north to Anzio, thus weakening the German line for a major Allied break through. Unfortunately, the American commanders were timid and did not seize the initiative. They thus failed to exploit their advantage over the temporary weakness in the German lines. Consequently, German forces were rapidly redeployed and the opportunity for a quick victory was lost. In fact, after four months, because of Lucas' nonaggressive posture, the Germans almost drove the American landing at Anzio back into the sea. Lucas was eventually relieved of command. After a four-month stalemate, the siege of Anzio finally ended on May 23, 1944, when the Allies launched a breakout offensive.

Meanwhile, in February, the Americans made several attempts to break through the Gustav Line northward at a place called Monte Cassino. As they sought to breach the defenses, Allied bombers demolished the ancient Roman Catholic monastery atop Monte Cassino, assuming it was being used by German artillery spotters. That assumption proved wrong, but the misguided attack reduced the monastery to rubble. This only made it more difficult to take when German snipers moved into the rubble as hiding places to fire at Allied forces. Consequently, the battle brought much controversy against the American bombing of the monastery.

Allied bombers thereafter found more appropriate German military targets units deployed lower on the mountain. Though they bombed it through February and March of 1944, the frequent air raids failed to destroy German defenses or enable a break in the German lines.

Finally, American deception succeeded where bombing had not. The US Army staged an elaborate faint, a fake attack, north of Rome forcing Albert Kesselring, the German commander, to redeploy forces from his defensive line at Cassino. Meanwhile, the US Army interdicted German supply lines impeding Kesselring from reinforcing his defenses at Cassino. By May, the Americans held a three-to-one advantage and broke through the depleted German line. They then rushed north liberating Rome and the territory in between. The Germans were again driven back, but not out of Italy. Kesselring, a master of defensive warfare, established another new defensive line, the Gothic Line, 200 miles north of Rome.

In the summer of 1944, the invasion of Italy ground to an anticlimactic end. Allied ground commanders believed they were poised to crush the new Gothic Line which stretched from Pisa to Florence. However, the high command in Washington and London rather determined that the Normandy invasion of France, of June 6, 1944, should be supported by an invasion of southern France. Therefore, divisions were transferred from Italy for that impending invasion in August of 1944. It took American and British forces until the end of the European theater of World War II to finally defeat the Germans in Italy. The Allies did not breach the Gothic line until the end of the war in April 1945.

The Italy campaign had been a bloody affair with over 119,000 American casualties. Of that number 59,000 died. The British suffered almost 90,000 casualties. Overall, the Allies including Canada, Australia, and other British dominion forces suffered around 330,000 casualties, about one half of their original strength. German casualties were estimated between 336,000 and 580,000.

As it turned out; the North Africa, Sicilian, and Italian campaigns of World War II were not decisive for the Allies. They did not directly lead to the invasion or defeat of Germany. In comparison to the magnitude of the Russian front and the coming invasion of Normandy, they were a back water of the war. Allied casualties were horrendous. However, there were great benefits nevertheless. The Allies were able to take complete control of the Mediterranean, wresting it from the Italian navy and the German Luftwaffe. It thus kept crucial supply lines to and from Gibraltar to Suez open for the British.

The Allies gained immense practical knowledge and experience in amphibious landings which would prove invaluable at the coming invasion of Normandy. Moreover, a large quantity of amphibious landing craft had already been built and battle tested. Furthermore, the green American army, in particular, gained vital combat experience for the coming battles in France. Senior American officers likewise gained invaluable combat experience which would prove crucial in what lay ahead. And, the battles of North Africa, Sicily, and Italy tied down many German divisions which could have been used profitably in Russia and later in France.

Out of these campaigns, British and American integrated leadership was knit together as a true Allied command. Few nations in history have ever enjoyed such long-term cooperation. It was forged in North Africa, Sicily, and Italy.

Finally, the occupation of Italy gave the American 15th Air Force strategic air bases in Italy for the bombing of the Third Reich. They in particular destroyed the German petroleum refineries and industries. That in itself was a major factor in the collapse of Hitler's Reich. And so, though the battles and victories in that region were not decisive insofar as final victory was concerned, nevertheless, the Sicilian and Italian campaigns played a very helpful role in the eventual Allied victory over Nazi Germany.

CHAPTER 25

Operation Overlord

Though there had been preliminary ground campaigns prior to the invasion of France—North Africa, Sicily, and Italy—the main ground war for American and British Allies began with what has ever since come to be known as D-Day. It officially was called Operation Overlord. This would be followed by almost a year of intense combat as Allied forces drove eastward to and finally into Germany. Overlord was the largest, amphibious, military operation in history to that time; though the American invasion of Okinawa about a year later was even larger. By the end of July 1944, two million Allied troops had been landed in France.

After the fall of France in the summer of 1940, along with the debacle of Dunkirk, British war planners along with the help of other Commonwealth countries and the United States, deemed it would not be possible to regain a foothold in continental Europe in the near future. After Hitler invaded Russia in June of 1941, Stalin began pressing the British, and then later that year the Americans, to open a second front in the west to relieve pressure on his faltering army. Several plans for invading France were considered in 1942-43, but were quickly shelved. Even with America now in the war, the Allies simply did not have enough strength at hand to successfully invade northwestern Europe. Nevertheless, the campaigns in Africa, Sicily, and Italy provided valuable experience for the looming battle in France.

British Chief-of-Staff General Frederick Morgan was ordered to begin planning an invasion of France in May of 1943, with a target date of about a year later—spring of 1944. Though the Pas-de-Calais area in France was the shortest distance from England across the English Channel, Morgan ruled it out for several reasons: (1) it would be the most obvious route of attack and (2) he knew the Germans were building heavy fortifications there. Other tactical issues also removed Calais from consideration. However, the region of Normandy further west offered much better potential. Normandy would allow simultaneous attacks against the port of Cherbourg and coastal ports further west in Brittany. It also would allow an overland attack towards Paris and eventually into Germany. Normandy was therefore chosen as the landing site. The one serious drawback of the Normandy coast was a lack of port facilities. However, that would be overcome through the construction of artificial harbors.

Subsequently, Dwight Eisenhower was appointed to be the commander of the Supreme Headquarters Allied Expeditionary Force (SHAEF). The Americans would supply the largest number of troops and equipment, therefore an American commander was chosen by Roosevelt and Churchill. British General Bernard Montgomery was named commander of the 21st Army Group, which would include all the land forces involved in the invasion.

Morgan had initially planned for an invasion force of three divisions, but when Eisenhower and Montgomery became involved in planning, they insisted on a minimum of five divisions. The invasion plan had initially been named Operation Roundup, but was later changed to Operation Overlord. On D-Day, June 6,

1944, about 160,000 Allied troops would be landed (and another three airborne division parachuted down), totaling 194,000 Allied troops.

Preparations for Overlord

In preparation for the European invasion, England became the largest military base in the world in preparation for the upcoming operation in France. Transport after transport of troops began arriving in England from the United States and British Commonwealth countries. Many American troops crossed the Atlantic in large passenger liners converted for troop transport. They were preferred because their high speed made them difficult targets for U-Boats. There were 37,000 US troops in England in June 1942. That grew to 790,000 at the end of 1943. By the time of the D-Day landings, there were over two million US personnel in England, not counting forces from Canada and other Commonwealth countries. Amazingly, no troop transports were sunk by German submarines. The Battle of the Atlantic had already been won.

The most of the troop concentration was in southwestern England. However, not only were troops arriving in England, but tens of thousands of Sherman tanks—eventually almost 50,000. Hundreds of thousands of 2 ½ ton American army trucks and jeeps arrived, not to mention tens of thousands of artillery pieces and their carriages. Then there was the endless quantities of ammunition, food, fuel, medical gear, communications equipment, combat engineering equipment, and general supplies arriving. Thousands of fighter bombers such as the P-47 and P-51 had to be shipped across the ocean aboard ships. Southwestern England became a huge military storage depot to the degree that locals joked about their island nation sinking into the ocean from the weight. The German's were aware of the buildup and that an invasion was coming, but they were in no position to do much about it. The air war over northwestern Europe had already been largely won by the Allies. The Luftwaffe was no longer a threat. The Kriegsmarine was basically in hiding.

As D-Day approached, virtually all of southern England was locked down. No one, soldiers or locals, was allowed to leave the region without military permission. Security was tight as a drum. As the day approached, hundreds of thousands of troops were in motion, moving to ports of embarkation for France. The massive amounts of tanks, trucks, and all the voluminous supplies were in motion, waiting to be loaded on the invasion transports. Thousands of ships were in English ports waiting to be loaded. If D-Day was the largest amphibious landing in history to that point, the logistics of preparation were also unrivaled.

Personnel were sealed into their marshaling areas at the end of May, with no further communication allowed with the outside world. Infantry units were briefed using maps which were correct in every detail except for the names of places, and most were not told their actual destination until they were already underway. A news blackout in Britain increased the effectiveness of the Allied deception plans.

Training exercises for the upcoming D-Day landings took place as early as July 1943. The town of Slapton in Devon, England, was evacuated in December 1943 and taken over by the armed forces because of its similarity to the Normandy region. In a training exercise there, a friendly fire incident on April 27, 1944 resulted in 450 deaths. The next day, an additional 750 American soldiers and sailors died when German E-boats (torpedo boats) attacked. News of that was censored. Training exercises with landing craft and live ammunition took place in Scotland. Medical teams in London and elsewhere rehearsed how to handle the expected volume of casualties. British and American paratroopers conducted rehearsals, including a huge exercise on March 23, 1944 observed by Churchill, Eisenhower, and other top officials.

Deception

In the months leading up to the invasion, the Allies conducted a number of operations to deceive Hitler of where and when they were going to attack. The overall plan was called Operation Bodyguard which was divided up into several subordinate deception operations. One was called Operation Fortitude North which was a campaign intended to deceive the Germans through a campaign of fake radio messages into thinking that the Allies intended to invade Norway. The purpose was to convince the Germans to maintain a substantial force in Norway in the event of such an attack. Of course, in so doing, there were that many fewer Germans forces which would be available to stop the Normandy landings.

A second and more substantial campaign of deception was named Operation Fortitude South. It was designed to deceive the Germans into thinking the Allies would invade in the Pas-de-Calais area of France in *July* 1944. A fictitious First US Army Group was created, supposedly located in southeastern England under the command of George Patton. Inflatable fake tanks, trucks, and landing craft were created and placed near the coast to hopefully be seen by German reconnaissance aircraft. Real military units, including several Canadian divisions were moved into the area to strengthen the illusion of a large force assembling there. In addition to fake radio traffic, genuine radio messages from real army units were first routed to the areas via land lines and then broadcast to cause the Germans to think most of the Allied troops were stationed there. Patton continued in England until July 6, deceiving the Germans into thinking a second attack would take place near Calais.

The Germans thought they had a substantial network of spies in England. However, every one of them had been caught by British intelligence and offered a simple decision: go to work for us or be hanged. Incredibly, every German spy in England, real or created by British intelligence, turned and worked for the British. They then were fed information to be radioed back to Germany. It contained just enough truth to convince the Germans, but crucial details were distorted. They were used to further the deception that the coming invasion would take place at Calais. Other agents across Europe, posing as Nazi spies but in reality working for the Brits, sent messages to Berlin, crafted by British intelligence, further strengthening the deception of an invasion at Calais.

Most of the German radar system on the coast of France had been destroyed not long before D-Day by the RAF. On the night before the invasion, Allied aircraft dropped aluminum foil strips called "Window" which deceived the remaining German radar operators into thinking a major naval force was approaching an area of the French coast about 60 miles east of Normandy and near the Calais area. The deception was strengthened by a group of small vessels towing barrage balloons. That same night dummy paratroopers were dropped over areas east of Normandy which led the Germans to believe an additional airborne assault was occurring there. All of these operation left the Germans not only deceived but confused. Without the extensive deception operations, the invasion very likely might have failed.

The Atlantic Wall

Adolf Hitler was well aware of the Allied build up in England. As early as 1942, he knew the Allies would attempt an invasion of what he called "Fortress Europe." On March 23, 1942 he therefore ordered, the construction of what he called the "Atlantic Wall" which he envisioned would withstand an invasion. It was an exercise in futility. Though one of the great engineering projects of the 20th century, it would fail in less than one day. His order directed the construction of 15,000 separate concrete gun emplacements to be manned by 300,000 soldiers. Because the Germans did not know where the invasion would take place, Hitler order his Atlantic Wall to run from Norway to the French-Spanish border. Depending on where one measures, it covered a distance between 1,670 to 2,000 miles. It was immensely expensive, costing Germany about $220 billion dollars in 2020 US dollars. Though Hitler wanted it done sooner, it took two years to build. The wall took more than 260,000 workers to build, though only about 10 percent of these men were German. The rest were either slave laborers or poorly paid local workers.

The first German commander ordered to oversee the fortifications was Field Marshall von Rundstedt, a general who had outflanked the Maginot Line during World War I. That maneuver ultimately precipitated the collapse of France. He now was ordered to build an even more massive line of fortifications. In late 1943, Field Marshal Erwin Rommel replaced Von Rundstedt and inspected the Atlantic Wall for the first time. He thought Hitler's wall was a gigantic farce. Rommel privately described Hitler's wall strategy as something from *wolkenkuckucksheim* or "cuckoo land." Nevertheless, Rommel's strategic preparations would ultimately enable the Germans to inflict terrible Allied casualties come D-Day.

In addition to literally millions of mines, thousands of machine gun nests, untold miles of barbed wire, there were thousands of concrete gun emplacements, many of large caliber. Those guns were a hodge-podge of sizes and calibers brought in from all over Europe. They ranged from naval guns cut out of old French and German warships, to tank turrets, to captured artillery pieces from Czechoslovakia and France. Maintaining and providing ammunition to this patch work of weaponry became a logistical nightmare for the Germans. Later when Rommel became involved with the wall, he added what came to be known as Rommel's asparagus which were sharpened poles placed in the ground to snare paratroops or gliders coming down. Endless anti-tank traps and mined obstacles were also installed at the water's' edge.

Perhaps, the heaviest fortifications were in the Calais area of France where the Germans thought the invasion was most probable. However, Rommel also realized that Normandy would be an ideal place for an amphibious invasion. He therefore ordered fortifications increased there. By the summer of 1944, German gun crews had spent months pre-sighting guns at anticipated landing beaches in Normandy. They constantly rehearsed placing artillery, mortar, and machinegun fire onto these pre-sighted killing zones.

Rommel made strategic proposals for defending the Atlantic Wall to Hitler. If Hitler had paid heed to Rommel, the D-Day invasion might have been a complete disaster for the Allies. In early 1944, Rommel wanted to place Germany's elite Panzer tank divisions as close to the French coastline as possible to repel an assault from the water. Hitler disagreed, choosing rather to position most of his Panzer divisions closer to Paris, and spreading the rest in southern France. They were to serve as a mobile reserve to be moved where needed when the invasion came. Rommel had only direct command of three Panzer tank divisions. Only one Panzer unit was relatively near the Normandy beaches. (It would wreak havoc on the British attempt to seize Caen after D-Day.)

Rommel was thus put in a strait jacket with rules which said he couldn't move his Panzer divisions to Normandy without Hitler's express approval. As it turned out, Hitler did not give that approval until it was too late.

However, the events on D-Day proved Rommel's fears. Hitler's Atlantic Wall was breached in less than a day on June 6, 1944. Part of the fallacy of Hitler's plan was trying to fortify up to 2,000 miles of the irregular European shoreline. He was spread too thin. A concentrated attack at any given point was likely to succeed, which it did that day. The attack at Normandy caught the Germans completely off guard. Though exact numbers of German defenders are not known, the attacking Allied forces at Normandy overwhelmed them that day, notwithstanding all the fortifications Hitler had installed.

Because of the elaborate Allied campaign of deception, Hitler remained sure the invasion at Normandy was only a feint and that the real attack would be at the Pas de Calais. Within days, the British, French, Americans, and Canadians had secured their beachheads through which millions of fresh troops would soon pour into Europe.

German Defenses

Though the Germans had 50 divisions in France and the Low Countries, along with additional divisions in Denmark not far away, they only had the German Seventh Army in the Normandy area. This included the German 352nd division at the Omaha beach area. It was comprised mainly of older men and conscripted soldiers from territories Germany had conquered in the east. They were equipped largely with inferior captured equipment and were not motorized. Later in the battle, the 12th SS Panzer Division Hitlerjugend (Hitler Youth) arrived, but they were mostly teenagers of the Hitler Youth movement. Hitler also had an entire army in the Calais area, but would not release it to Normandy because he was certain an even larger invasion was coming there. Though under the expert command of Erwin Rommel, the Germans in Normandy were hampered by several significant factors.

First, because of the intensity of low-level bombing and strafing by the American Ninth Air Force, railroads, bridges, and highways in northern France had been largely wrecked and were under constant attack. Once the battle had been joined in Normandy, the Wehrmacht had severe problems moving reinforcements to the front. They were reduced to marching only at night, and on foot for the most part. No longer could the Germans rely on highly mobile, mechanized forces to which they were accustomed.

Second, because the battles in Russia to the east were intensifying, the German High Command was forced to transfer the entire II SS Panzer Corps from France to the eastern front. This took place before the Normandy landing. In short, the Wehrmacht in northern France was not at full strength. Other German divisions had been relocated to France in the Spring of 1944 from the Russian front to rest and refit, but they were not fully operational or mobile come June of 1944.

Invasion Date and the Weather

Because the invasion of France would be waterborne; weather, meteorological, and sea conditions were critical. The Allied planners considered a full moon to be advantageous because it gave some illumination for transport aircraft with airborne troops. A full moon also meant highest tides. The planners wanted the landings to be shortly before dawn, midway between low and high tide, with the tide coming in. This would help landing forces to better see the obstacles the Germans had placed in the water and along the beaches as they came ashore. There were also specific criteria for wind speed and direction which translated into sea conditions, along with visibility requirements and cloud cover. High winds meant high seas which would make loading troops onto landing craft and coming ashore perilous if not impossible. Cloud cover would hamper air operations.

The initial date when most of these criteria intersected would be on June 5th. However, on June 4th there were high winds, heavy seas running, and driving rain. The weather was absolutely awful that day. All the complex and intricate plans for embarkation of troops, ships putting to sea, and air operations were synchronized for the attack to begin on June 5th. Upwards of 194,000 soldiers were primed and ready to go, not to mention tens of thousands of sailors and airmen. Some men had to board their craft nearly a week before departure.

D-Day, incidentally, was simply military jargon for "the day," the *day* when the invasion would take place. But D-Day would have to wait.

On the evening of June 4th, however, the Allied meteorological team advised conditions would improve sufficiently so that the operation could take place on June 6th. Otherwise, the soonest tides would be suitable again would be June 18th and a full moon would not appear again until July 3rd. With troops already boarded and the great operation ready to go like a giant compressed spring, Eisenhower gave the command, "Okay, we'll go," the night of June 4th for the attack to begin on June 6th. And so the orders went out and the greatest amphibious attack in history to that point, like a gigantic machine began to move. Though the weather was considered adequate, it was only marginally so. Choppy seas would cause seasickness and conditions rough enough, particularly off Omaha Beach, that specialized swimming tanks, for example, simply sank, depriving the infantry of the planned armor they were expecting.

CHAPTER 26

The Invasion Begins

I n the days prior to June 6th, Eisenhower, issued this letter to the troops. It read:

> "You are about to embark upon the Great Crusade, toward which we have striven these many months. The eyes of the world are upon you. The hopes and prayers of liberty-loving people everywhere march with you. In company with our brave Allies and brothers-in-arms on other Fronts, you will bring about the destruction of the German war machine, the elimination of Nazi tyranny over the oppressed peoples of Europe, and security for ourselves in a free world."

On the evening of June 5th, minesweepers began clearing lanes for the convoys which would cross the English Channel during the night. Just before dawn, approximately 1,000 Allied bombers swept over Normandy and bombed what they hoped were the German defensive positions below. Once again, precision bombing, or the lack thereof, became an issue. Though dropping thousands of tons of bombs along the beachheads, they largely missed the crucial German coastal positions. In addition to neutralizing

German gun emplacements, the Allies hoped the bombers would also hit the beaches, leaving bomb craters in which landing troops could take cover from enemy fire. Unfortunately, the bombs fell far enough inland that neither German coastal positions nor the beaches were hit.

Just before midnight on June 5th, twelve-hundred C-47s or similar transport aircraft departed for Normandy, many towing assault gliders. Three airborne divisions were on board headed for drop zones behind German lines to land that night. On the western end of the assault beaches, the American 82nd and 101st Airborne Divisions landed near the Cotentin Peninsula. On the eastern end of the assault beaches, sixty miles away, the British 6th Airborne Division was dropped to capture intact the bridges over the Caen Canal and River Orne.

The Germans had deliberately flooded the area at the base of the Cotentin Peninsula where the American paratroops arrived. They accordingly suffered substantial casualties by drowning but nevertheless secured their objective. The British 6th Airborne Division seized its un-flooded objectives at the eastern end of the assault beaches more easily and captured the key bridges over the Caen Canal and Orne River.

At 5:45 to 6:25 that morning, Allied warships offshore began a naval bombardment of German coastal defenses along the assault beaches. Included were five battleships, twenty cruisers, and sixty-five destroyers. The first infantry began arriving on the beaches at about 6:30 a.m. At Normandy, approximately 6,000 vessels were involved, from warships to troop transports, to landing crafts of various sizes.

The Beaches

The Normandy assault beaches covered a distance of about 60 miles. It was divided into five specific zones on which the allied divisions would land. From west to east, the five beaches were Utah, Omaha, Gold, Juno, and Sword. The Americans were assigned to Utah and Omaha. The British and Canadians were assigned Gold, Juno, and Sword beaches. Because of tidal action, the American landing at Utah was first at about 6:30 am, and then progressing easterly over the next hour: Omaha, Gold, Juno, and Sword. One objective was to catch the Germans off guard. In that regard, the Allies largely succeeded.

Utah

American assault units landed on Utah beach to relatively light resistance. Though their landing craft had been driven about a mile east of their intended landing zone by wind and tidal currents, they were able to regroup and begin moving off the beach, suffering less than 200 casualties. Their efforts to move inland was short of their objectives for the first day, but they were able to advance about four miles and link up with remnants of the 101st Airborne Division. The airborne landings in that area had been scattered, largely due to pilot error and only about 10 percent landed on target. The rest were scattered across the terrain with its hedgerows, stone walls, and marshes. The 82nd Airborne achieved its primary objective at Sainte Mère Église and worked to protect the western flank of the invasion area. Utah beach would prove to be a relatively easy objective, compared to what took place at Omaha beach.

Omaha Beach

Omaha was ghastly different. The coastal topography was of cliffs and bluffs atop which the Germans had placed deadly installations of interlocking machine gun fire and artillery—zeroed in on the beach below. There were also major obstacles installed at the water's edge. Unknown to the Americans, the Wehrmacht had recently moved their 352nd infantry division into that area, rather than just a regiment the Americans were expecting to face. The Germans are thought to have had perhaps 7,800 men atop the bluffs and cliffs overlooking Omaha. But they were strategically located in such a way that their relatively low numbers were able to inflict horrendous losses on the Americans below. The Germans had pre-sighted 85 machine gun emplacements in addition to 35 other pillboxes with at least a dozen artillery sites.

Strong tidal currents swept landing craft east of their intended objectives, causing confusion for the troops who managed to land. The Germans at Omaha quickly perceived what was taking place and opened up with everything they had on the hapless assault forces. It became a killing field. Some landing craft did not navigate through the mined-tipped obstacles and sank before they reached the beach. Most other landing craft were greeted with fusillades of accurate machine gun fire as they dropped their landing ramps and began off-loading troops. Many men were shot dead immediately after jumping into the water—many were shot dead before they even left the landing craft. Some landing craft dropped their ramps in water too deep, drowning the heavily laden infantrymen as they dropped into water over their heads. Those who made it ashore were mowed down like grass.

Plans were for the dual-drive (DD) tanks to swim ashore with a temporary canvass shroud giving them buoyancy. However, (1) their landing craft launched them too far out to sea fearing shore artillery. (2) Because seas were choppy-to-rough, almost all of them were swamped and sank. The upshot of it was the hoped-for armored support on the beach for the American troops never materialized, leaving them at the mercy of deadly German gunfire. Just about everything that could go wrong did. Soldiers who did make it ashore cowered and hid behind beach obstacles or anything that would be a shield from the deadly enfilading fire. Even the dead bodies of fallen comrades were used for shelter.

However, little by little, some began to sprint across the beach to shelter at the base of the bluffs. Meanwhile, a group of allied naval destroyers came as close in as their draft allowed and began to fire on the sources of German fire. That helped tremendously. By noon, as the naval guns took its toll and as the Germans started to run out of ammunition, the Americans were able to open lanes across the beaches.

Men on the beaches came to the conclusion if they stayed on the beach, cowering behind whatever, they would die. Therefore, some seized the initiative and headed for cover below the cliffs and bluffs. Exits from Omaha were six gullies or draws reaching up away from the beach. They, of course, had been heavily fortified with barbed wire, concrete obstacles, and mines. Small groups of American soldiers began to clear the draws and gullies and by mid-day American forces were reaching the top of the cliffs and bluffs. This allowed military vehicles to begin to work their way inland. As the day progressed, little by little, more and more troops were coming ashore with more and more equipment. The crisis had past. They began, albeit slowly, to move off the beaches in force and seize the high ground from the Germans.

On the border between Utah and Omaha beaches was a point of land named Pointe du Hoc. There, Allied intelligence believed large caliber German coastal defense guns were emplaced in a heavy concrete casement. The apparent danger was the potential for them to sink ships lying offshore. The 2nd Ranger Battalion of the US Army was charged with destroying the big guns. They were forced to scale a 100 foot high cliff using ropes and rickety assault ladders. It proved as difficult as it might sound, but they, one by one, made it to the top and assaulted the gun position, only to find it armed with telephone poles. From aerial photography, they had looked like real artillery. The rangers found the big guns about a third of a mile inland and disabled them with white-phosphorous grenades which melted their firing mechanisms, rendering them useless.

At one point, the slaughter at Omaha seemed so bad, commanders aboard ships offshore contemplated cancelling the operation. However, they did not. The Americans suffered 2,400 men dead at Omaha on June 6, but by the end of the day they had landed 34,000 troops there. The German 352nd Division opposing them lost 20 percent of its strength. However, they had no reserves coming up to continue the fight. They also were virtually out of ammunition with little prospect of resupply. The US Army suffered grievous losses at Omaha; but it had landed, established a beachhead, and was moving inland. Though often remembered for the terrible initial losses; in the greater picture, the landing there succeeded. As the other beaches were assaulted, Hitler's Atlantic wall was breached in less than a day.

Gold Beach

Just to the east of Omaha, British and Canadian forces were scheduled to assault their beaches. The wind had piped up making it difficult for landing craft. Therefore, the DD tanks (double drive swimming tanks) were landed close to shore or directly onto the beach. One target of the preliminary bombing earlier in the morning had been a 75 mm gun emplacement. The bombers missed, allowing the gun to do much damage until about 4 p.m. in the afternoon, when it was finally silenced. A British battalion captured Arromanches which would become the site of one of the artificial Mulberry harbors. Total British-Canadian casualties at Gold were approximately 1,000, of which 350 were killed. German losses are unknown; at least 1,000 were taken prisoner. Initial objectives were taken.

Juno Beach

Northwesterly winds continued to build as the sun rose causing rough seas at Juno beach. The landing was delayed somewhat, but infantry finally landed ahead of their scheduled armor (i.e., tanks) resulting in unnecessary casualties. Preliminary bombing, once again had missed their targets as had much of the naval bombardment. Nevertheless, Canadian units quickly cleared the beach and constructed two exits to the French villages above. By evening, Juno and Gold beachheads covered an area 12 miles long and 7 miles deep. The Canadian Army lost 340 killed, 574 wounded, and 47 taken prisoner for a total of 961.

Sword Beach

At 7:30 a.m., British forces landed on Sword Beach. They were able to successfully land 21 of 25 DD tanks (i.e., swimming tanks). These immediately began providing cover for infantry troop landing around them. (If the American DD tanks had been able to get on the beach at Omaha, it would have been a different story there.) Combat engineers were able to quickly open several exits off the beaches for the tanks, allowing troops to begin moving inland. At 4 p.m. that afternoon, the Germans counterattacked between Sword and Juno and nearly succeeded in reaching the coast. However, they met fierce resistance from the

British and retreated to assist other German forces in the area near Caen. The British suffered 683 casualties at Sword.

Aftermath and Overview

The Allies suffered around 10,000 casualties with 4,414 dead on the first day. However, in perspective, around 160,000 Allied troops were landed over the beaches on D-Day. Allied planners had expected far worse losses. A beachhead had been established and the Germans, though fighting fiercely in retreat, did not have the strength or reinforcements to repulse the Allied forces.

The Allied invasion objectives on the first day were for the capture of French towns of Carentan and St. Lô by the Americans, with Caen and Bayeux taken by the Brits and Canadians. The plan was for all the beaches (other than Utah) to be linked into a front line 6 to 10 miles from the beaches. None of these goals were achieved the first day. All five beachheads were not connected until June 12th. By then the Allies held a front approximately 60 miles long and 15 miles in depth.

The Americans had intended to take the port of Cherbourg as soon as possible, but terrain behind the American beaches was the bocage, with its thick hedgerows. Many areas there were additionally protected by rifle pits and machine-gun emplacements. Most of the lanes were too narrow for tanks and the Germans had flooded the fields behind Utah up to two miles from the coast. By D-Day plus 3, the Americans realized they could not quickly take Cherbourg and rather endeavored to cut off the Cotentin Peninsula of which Cherbourg was at the top. Cherbourg was not captured until June 26th, but by then the Germans had utterly destroyed the port facilities making the harbor useless for several months.

On the eastern end of the D-Day beaches, the British objective for D-Day was to capture the French city of Caen, a strategic road and rail center. It was located about ten miles south of the Sword beach. However, fierce German resistance along with sluggish action by British General Montgomery caused battles to rage over Caen well into July. Caen, a major objective of the Brits and Canadians, remained in German hands at the end of D-Day and would not be completely captured until July 21st. British bombing of the city and a final all-out assault then enabled British and Canadian forces to seize it.

Mulberries

The Allies knew they would desperately need port facilities to pour many more troops into France after D-Day and keep them supplied. The Americans had hoped to quickly capture the port of Cherbourg just north of the Utah beach area. However, it took them almost three weeks to accomplish that. During that time, the Germans had completely wrecked the port facilities.

Anticipating that, the English had devised an ingenious system of artificial harbors called Mulberries. They were a system of concrete caissons which were designed to initially float. They were towed across the English Channel on the day after D-Day (D+1) and then sunk at predetermined sites just off the beaches of Normandy. Two Mulberry systems were installed: one for the British and one for the Americans. The mulberries were fortified with old ships which were sunk as part of a breakwater system a bit further out. The actual unloading facilities were designed to float up and down with the tide. This was all connected to the shore by a series of floating, flexible steel bridge sections, six miles long. The system worked remarkably well—at least for a while. They were designed to handle up to 7,000 tons of cargo and trucks daily. Within twelve days, the artificial harbors were operational.

But alas, Mother Nature ruled the day. On June 19 a violent storm began and by June 22 the American harbor was destroyed. However, the Americans had designed and built many landing ships called LSTs (Landing Ship Tank) which could drive right up on the beach, open massive front doors, and offload heavy equipment such as tanks, bulldozers, trucks, and all the necessary equipment needed by the US Army. They were slower, simply because there were not enough of them, but they worked and the LSTs would continue to supply American forces for much of the summer of 1944 until the Cherbourg harbor was repaired and rebuilt. The Normandy harbors would receive about 2.5 million men and 500,000 vehicles throughout the remainder of the war.

PLUTO

One of the crucial needs of a modern mechanized army is fuel—gasoline in particular. Knowing Allied armies would be thirsty for fuel daily, the Allies designed an undersea pipeline from England to Normandy. It was called PLUTO—Pipe Line Under The Ocean. Work began on it not long after D-Day and by September it was operational. It would eventually pump over 300 tons of fuel a day to Normandy where it was then distributed to countless allied tanks and vehicles.

Hitler's Reaction

As the Allies came ashore on the Normandy beaches D-Day morning, Hitler was sound asleep back in Germany. His subordinates had given strict orders the Fuhrer was not to be awakened. Notwithstanding, his phone began to ring. Field Marshal von

Rundstedt, the top German commander in France desperately wanted instructions and specifically for Hitler to release the two Panzer divisions which were in the vicinity of Paris. The Fuhrer's staff refused to awaken him. When he finally got up around noon and was informed of what was taking place, Hitler refused to act. He was convinced what was happening at Normandy was a ruse by the Allies and the real attack was yet to occur. When it did, Hitler thought it would be led by American General George Patton at Calais. Hitler had swallowed the Allied deception plan hook, line, and sinker. Furthermore, OKW (the German overall military headquarters) was also reluctant to authorize their release—at first. Their reaction was, what if the "invasion" was a trap?

Finally, after lunch, Hitler reluctantly agreed to von Rundstedt's request. But by then it was too late. If the Panzer units had moved out early in the morning, under cover of darkness, they might have reached the front. But now they would have to wait out the daylight hours. The American Ninth Air Force swarmed the skies over northern France and their deadly fighter bombers would decimate any such a move in daylight. Meanwhile, Allied forces at Normandy were pushing inland by noon. By the time German reinforcements would have arrived, the Allies had secured their beachheads. By the end of the day, there were about 160,000 Allied troops in Normandy, with hundreds of thousands more arriving in the next several days. They already greatly outnumber German forces in the area, actual and potential.

Stalemate

The planners of Operation Overlord had been thorough and meticulous in their planning. Though there were difficulties, particularly at Omaha Beach, overall the invasion had gone well from its preparations, to its logistics, to its execution. However, it seems that the planning had extended only to the arrival on the beaches. Thereafter, things did not go well. For the next seven weeks, Allied forces made little progress. July of 1944 proved to be an anxious time for the Allied leadership as the Allied force bogged down in Normandy. By July 1, 1944, the Allied beachhead was only 20% the size Overlord's planners had hoped it would be one month after D-Day. Furthermore, Allied commanders were increasingly worried the Germans might actually commit their reserves to Normandy. Hitler still clung to his view that a larger invasion was going to take place at Calais, though that conviction was waning. If Hitler and the German high command released their 15th Army at Calais, the Germans might have indefinitely tied down the Allies in Normandy. Visions of a return to the horrors of World War I trench warfare gripped the Allied command.

On the western portion of the invasion beachhead was the bocage country of the Cotentin peninsula. Over the centuries, French farmers had built substantial hedgerows around their fields which were called the bocage. They had excavated sunken lanes between fields and used the excavated dirt to build up mounds on each side from which grew up thick bushes and trees. They were almost like walls along the lanes. The entire region was an irregular quilt-work, checkerboard of these fields and pastures.

The Germans found the bocage to be ideal defensively. Machine gun emplacements were sited where they could effectively hit Americans soldiers trying to breach any given hedgerow. American Sherman tanks could barely climb over them. But as they did, their un-armored undersides were completely exposed making them a perfect target for anti-tank guns. American advance each day was measured in yards and was bloody at that.

The American planners of the Normandy campaign surely had seen aerial photographs of the bocage country, but apparently did not realize how substantial the hedgerows were. No American operation had ever tried to advance through such country. Various techniques were experimented with, though mostly unsuccessful. Finally, American enlisted men used scrap steel removed from German beach obstacles and fashioned them into sharp teeth, welded onto the front of Sherman tanks. They were nicknamed "Rhino tanks." They could cut their way through the hedgerows without exposing their soft under bellies and could then clear a given pasture of German machine gun emplacements, allowing infantry to secure each field. Little by little, the US Army was able to begin to make substantial movement through the bocage region.

The French town of St. Lo, a strategic crossroads at the base of the Cotentin peninsula, had been an American objective from D-Day itself. But it took American forces until July 7-19 to capture what was left of the bombed out town. Five American divisions sustained 11,000 casualties during the almost two weeks it took to take the city. The Americans were advancing, but slowly and with considerable casualties.

On the eastern end of the Normandy beachhead, the British were having their own troubles. Their objective from D-Day had been to capture the French city of Caen, an important road and rail junction. The German high command viewed Caen as a strategic hinge from which the British could pivot eastward and drive straight toward the German heartland. Therefore, they concentrated strong forces, including several Panzer divisions in and around Caen to prevent it from falling. The British faced a veritable defensive wall at Caen. It soon turned into a stalemate—an irresistible force against an immoveable object. Fierce fighting continued for almost a month at Caen. In late June, the British made an attack, swinging around to the southwest of the city, destroying a large number of German tanks, after the Germans had committed every available Panzer unit to the battle. The British operation failed, but the German commander for France, von Rundstedt, in despair made the remark that the war was lost. Rundstedt was thereupon relieved on July 1st and replaced by Field Marshal Günther von Kluge. Through British bombing of the German positions, Caen eventually fell on July 21. Both sides were exhausted.

Meanwhile, American First Army commander, General Omar Bradley, had set about planning a breakthrough of the German defenses just south of St. Lo to the west. Doing so would remove the last barrier between the Allies and the relatively open tank country of central France. He determined to leave nothing to chance.

Operation Cobra

With Operation Cobra, the stalemate at Normandy would be broken and American forces in particular would break out into central France. The Germans were occupied with the battles around Caen to the east. Bradley therefore devised a powerful plan whereby the American First Army would break through German defenses south of St. Lo while the Germans were off balance and had their hands full at Caen. Once a corridor had been opened, American forces could sweep into Brittany and thence into the interior of France.

After getting off to a slow start, the operation gathered momentum and German resistance collapsed. Scattered remnants of broken units sought to escape to the Seine River. Without reinforcements to deal with the situation, the German response was weak and the entire Normandy front quickly collapsed. Operation Cobra, working in parallel with the British Second Army and the

Canadian First Army at Caen, enabled a major break out of Allied forces from Normandy.

The operation began on July 25th. That morning at about 9:30 a.m., 600 fighter bombers of the American Ninth Air Force attacked German defensive and artillery positions in an area only 300 yards wide along the German line in the St. Lo area. An hour later, 1,800 heavy bombers of the US Eighth Air Force began saturation bombing of German lines in an area 3.4 miles by 1.3 miles on the Saint Lô-Periers road. This was followed by a third and final wave of B-25 and B-26 medium bombers. Approximately 3,000 American aircraft carpetbombed this narrow section of the German front.

At the outset of Operation Cobra, the German Panzer Lehr Division had only 2,200 combat troops, 12 Panzer IV tanks and 16 Panther tanks to start with. It was in the path of the Allied saturation bombing. When it was over, the division suffered about 1,000 casualties. Exhausted and demoralized, its commander reported his Panzer division was "annihilated," with its tanks wiped out, its personnel either casualties or missing, and all headquarters records lost.

The Germans who survived were dazed. Tanks were upside down. The area was pulverized. By 11 a.m. that morning, American ground forces began to pour through the gap in the German line. As they passed the area of the bombing, German resistance increased and the battle was joined. But the back of the German front south of St. Lo had been broken. Meanwhile, the German high command remained convinced their real threat was still at Caen and did not send reinforcements to the St. Lo area. Nevertheless, German resistance remained stiff and the American First Army had its hands full. But by July 27th, American forces began to break through German lines and were advancing rapidly. By July 28th, German defenses along the US front had largely collapsed and resistance was disorganized and sporadic. That day, P-47 Thunderbolts destroyed 122 German tanks, 259 other vehicles, and 11 artillery pieces.

When German commander Guderian was ordered to concentrate his division for a counterattack against the surging Americans, in frustration he replied, his panzergrenadiers could not succeed because of the intensity of Allied air attacks, the P-47s in particular. Meanwhile, advancing southward along the French coast on July 30th, American forces seized the town of Avranches, described as the gateway to Brittany and southern Normandy. By the July 31st, the US Army had repelled the last German counterattacks after fierce fighting, inflicting heavy losses on the Germans. The US advance was now unstoppable and the First Army was finally free of the bocage.

Our military genius, Herr Hitler, was in fact showing himself again to be a dummkopf. His catastrophic blunder in declaring war on the United States was coming home to roost. The Americans were now poised to race across France toward Germany. Meanwhile, the Russians were about to begin their final drive against the German homeland. It was called Operation Bagration. The vise was beginning to tighten.

CHAPTER 27

Operation Bagration

A fter the major battles of Stalingrad and Kursk in Russia in 1943, the Red Army spent the late winter and spring months of 1944 refitting, regrouping, and cleaning up their flanks. However, come summer of 1944, about the same time as the Normandy Invasion, the Soviet Union would unleash what many military historians consider to be the greatest and most significant campaign of World War II. It was called Operation Bagration, named after Pyotr Bagration, a Russian hero in the war of 1812 against Napoleon.

Virtually everyone in the United States has heard of D-day and perhaps the Battle of the Bulge. Few have ever heard of Operation Bagration. It was, arguably, the single most successful military action of the entire war, destroying the heart of the German army in Russia. In area, Bagration dwarfed the campaign for Normandy. In one month, it inflicted greater losses than the Wehrmacht had suffered in five months at Stalingrad. With more than 2.3 million men, Bagration was the largest *Allied* operation of World War II. It destroyed three German armies and tore open the Eastern Front. Operation Bagration was designed to break the back of the Wehrmacht once and for all. The sheer magnitude of it completely overshadowed Operation Overlord and D-Day.

The Red Army destroyed 31 of 34 divisions of Hitler's Army Group Center and completely shattered the German front line. It was the greatest defeat in German military history, killing around 450,000 Axis soldiers, while 300,000 others were cut off and became prisoners of war.

Whereas in 1941, the Red Army had resorted to crude, blunt-force, human-wave tactics to survive, by 1944 they had become more sophisticated. The Russian general staff, called Stavka, had learned from their own mistakes and from the tactical skill of the Germans. They likewise had learned lessons from the British and Americans in using deception to deceive and mislead Hitler.

The eastern front in the spring of 1944 found the Wehrmacht in similar formation as it had been in 1941. To the north, German Army group North under General Georg Lindemann had been dislodged from Leningrad, freeing the city from its deadly siege. Yet, it remained a strong German force in northwestern Russia. After the Russian success in retaking most of the Ukraine in 1943, to the south was German Army Group South. Hitler *re-named* it as Army Group North-Ukraine under Field Marshall Walter Model. In the middle was Army Group Center under Field Marshall Ernst Busch.

Operation Bagration would target Army Group Center which largely was in Belorussia (also called White Russia and today is called Belarus). There, the main Soviet attack would fall. Hitler had ordered 34 infantry divisions to Army Group Center in Belorussia. He also sent other miscellaneous divisions for a total of 50 German divisions. However, he soon would divert 16 of those divisions southward. These all were organized into four armies and commanded by Field Marshal Ernst Busch, whose appointment was mainly because of his loyalty to the Fuhrer. Army Group Center was in fact the anchor of the whole eastern front, blocking the shortest path to Berlin. The Russians would destroy it at the same time the western Allies were landing on D-Day, liberating Paris, and then driving towards Germany. Herr Hitler was in trouble, though he would not admit it or perhaps did not even realize it.

The Red Army planned a sophisticated operation against Hitler's armies. The German high command (OKW) anticipated the Russians would attack in the summer of 1944, but thought they would drive to the southwest against Army Group North-Ukraine. The Germans presumed the Soviet's ultimate objective was to move into Romania and seize the oil fields near Ploiesti. (The Red Army would ultimately do that, but that was not their immediate objective in Bagration.) Therefore, the Russians embarked on a campaign of deception, not unlike that which the British and Americans had done leading up to the Normandy Campaign.

Deception

The word the Russians used for deception was *maskirovka* which is roughly equivalent to the English word *camouflage*. It was their term for a campaign of deception against the Wehrmacht. The Red Army wanted the Germans to *think* they were planning to attack southwestward. The Russians kept substantial divisions in that region and made sure the Germans knew it. The Russian air force allowed just enough German reconnaissance aircraft to get through and see the Russian formations in northern Ukraine. Because the Russians were operating on their own soil, they sent communications back and forth to their armies in the field by secure telephone lines. However, they openly broadcast orders by radio designed to make the Germans think large Russian forces were assembling in northwestern Ukraine. The deception campaign was very similar to Operation Fortitude which the Western

Allies had used to convince Hitler the invasion of France would take place at Calais.

The Germans therefore began shifting divisions away from Army Group Center to Army Group North-Ukraine. Hitler was thus convinced the next attack by the Red Army would be launched in the northern part of the Ukraine. He therefore ordered reinforcements for the eastern front be diverted to Model's Army Group North-Ukraine, leaving Busch's Army Group Center with only 11 percent of the tanks and assault guns allocated to the Eastern Front. While some officers of Busch's intelligence staff predicted a major Belorussian offensive in late June, Busch himself accepted Hitler's assessment as more accurate. Following orders from OKW to the letter, he refused to let his commanders pull back or shorten their defensive lines more tightly in anticipation of an attack.

Since the German invasion of Russia in 1941, partisan guerilla groups of defiant Russian citizens had worked behind the front to sabotage German supply lines and cause damage in German rear areas. Though appreciated early in the war by the Russian high command (i.e., Stavka), there had been little effort to coordinate or supply partisan efforts. Now, Stavka actively began to coordinate and integrate partisan action into their greater strategic plans. The start of Bagration involved many partisan formations in Belorussia which were directed to resume attacks on railways and German supply lines. From June 19th onward, large numbers of explosive charges were placed on railroad tracks causing significant disruption, detonating some 10,500 demolition charges during the night of June 19-20 alone. This interrupted the movement of German ammunition, food, and reinforcements to the front.

Another problem for the German commander of Army Group Center was that his army, while strong in raw numbers, included a significant proportion of Luftwaffe field units and security troops—not trained infantry units. There were additional questionable Hungarian-Slovak divisions and also Volksdeutsche. The latter were ethnic Germans from occupied territories whose willingness to lay down their lives for the Fuhrer was suspect. Worse yet was the weakness of the Luftwaffe in the area. Germany's Sixth Air Fleet was vastly outnumbered by the Russian Air Force along Army Group Center's front.

From early when the Americans entered into the war, the Lend-Lease plan had, among other things, supplied the Russians with over 400,000 two-and-one-half-ton army trucks. These were integrated into the Bagration operation. As the attack against Army group center was about to commence, the Russians moved the several army groups which had been used as decoys in their deception plan. These were rapidly transported by night in the American trucks to useful positions before the attack began.

The Russians had also learned the hard way that Communist Party Commissars, assigned to each military unit in the Red Army to ensure compliance with the party line, were a hindrance militarily. By the time of Bagration, that silly system was dropped. Additionally, their already potent T-34 tank had been improved and up-gunned. New T-34s were equipped with 85mm guns which were now a match for German Panther tanks. They also were equipped with wider treads which would be more useful when crossing the marshy areas of western Russia and East Prussia.

In contrast to the practice, early in the war, when Russian commander Zhukov marching infantry directly into mine fields and sacrificing those poor soldiers, the Russians had now developed mine detonating tanks. These tanks had heavy drum-rollers mounted out in front of the tank, which when rolled over mines, detonated them ahead of the tank. This easily cleared mine fields, saving untold lives of Russian soldiers.

The Attack

The Russian attack began on June 22, 1944, exactly three years to the day when Hitler invaded Russia. The Red Army offensive was a characteristic Soviet enterprise—a massive push along a 450-mile-long line of advance. Four army group fronts would launch artillery barrages and attack simultaneously.

On the northern portion of the Russian line, the First Baltic Front under General Ivan Bagramyan fielded 359,500 men. They would push into Latvia to screen the right flank of the main assault from Hitler's Army Group North as well as to support forces farther south. Just to the south, the Third Belorussian Front under General Ivan Chernyakhovsky had 579,300 men, attacking heavily defended Vitebsk and the area north of city of Orsha. They then pushed southwest toward Minsk, the Belorussian capital, and Vilnius, the Lithuanian capital. They thereupon crushed or encircled Busch's Third Panzer Army at Vitebsk and his Fourth Army, centered around Orsha. South of Orsha, General Georgy Zakharov's Second Belorussian Front, attacked with 319,500 men. They helped complete the encirclement of Minsk and then the push westward.

Farthest to the south, the First Belorussian Front fielded 1,071,100 men commanded by General Konstantin Rokossovsky. They assaulted Busch's Ninth Army, skirting the Pripyat Marshes and pushing due west in the general direction of Minsk. The First and Third Belorussian fronts, which held the bulk of Russian armor and firepower, attacked along converging lines with the goal of encircling the German armies east of Minsk. To aid the attack, partisan units coordinated by Stavka launched demolition attacks against Belorussian railways preventing reinforcements from reaching the German front.

Inasmuch as Operation Bagration was immense and complex, the four Russian army group fronts were under the overall command of two trusted Stavka general officers. Marshal Aleksandr Vasilevsky, the organizer of victory at Stalingrad, commanded the two northern fronts, while the southern fronts were commanded by Marshal Georgi Zhukov.

The Red Army assembled 118 rifle divisions, eight tank and mechanized corps, thirteen artillery divisions and six cavalry divisions, with a total of approximately 2.3 million troops. The attack into battle was led by the rifle and tank divisions, which collectively fielded 2,715 tanks and 1,355 assault guns. To supply the offensive, the Red Army had stockpiled 1.3 million tons of ammunition, rations and supplies behind the front lines.

Ground troops were supported by 10,563 heavy artillery pieces along with 2,306 Katyusha multiple rocket launchers, nicknamed "Stalin's Organ" because of their pipeorgan appearance. Air cover was provided by 2,318 fighters of various types, 1,744 Ilyushin Il2 Shturmovik groundattack aircraft, 655 medium bombers and 431 night bombers. Another 1,007 medium bombers were drawn from the Soviet strategic bomber reserve. This totaled over 6,000 Russian aircraft in support of Bagration.

Operation Bagration thus launched, June 22, 1944, at 5 a.m. with a massive artillery bombardment. Each of the thousands of artillery pieces along the line was allocated approximately six tons of ammunition to fire during the two-hour barrage. The shelling was conducted in a rolling manner intended to destroy the

Wehrmacht's forward positions, then catch retreating soldiers in the open before they could reach safety to their rear. The less precise Katyusha batteries rained down on targets with 82mm and 132mm rockets ensuring nothing remained alive in the targeted positions. Stunned German survivors would describe this barrage as the most intense and terrifying barrage they had ever experienced.

The next day, June 23rd, the full force of the offensive launched forward. Abandoning their costly human-wave tactics of 1941, the Red Army concentrated its fire against tactically important targets, seized them, and then called up tanks to the new positions to deliver a greater breakthrough. By the afternoon of the seconD-Day, the German's Third Panzer Army's line had numerous holes in it and the German-held city of Vitebsk was in danger of encirclement by two Russian armies.

As the Russian Forty-third Army closed in around Vitebsk from the north and the Thirty-ninth Army attacked from the south, Busch sought permission from Hitler to withdraw to a secondary line of defense. But Hitler, thinking the main blow would fall elsewhere, had designated Vitebsk a "fortified place," to be held to the last man. By nightfall, two German divisions were encircled and two others were fighting for their lives. Once again Hitler blundered to the detriment of his own armies.

Hitler's order of March 8, 1944 had announced that *feste Plaetze* (fortified places) would be the core of the German defenses. The idea was that the Russians would not advance past these fortifications, and they would fulfill the function of fortresses as in earlier historical times. A city designated as a fortified place therefore could not surrender but must fight to the last man.

Hitler's orders to not retreat were a total disaster—another blunder. Refusing to allow his generals any flexibility or leeway was contrary to all the principles and the tenets of the German general staff. Hitler distrusted his generals because he thought them weak and lacking courage. He therefore demanded control of even local tactical decisions. This was basically the undoing of the German army. A Russian officer in the Bagration operation after the war said that the German behavior in their fortified places was "stupid." The fortified areas were smashed completely. Many Germans held their ground to the last man—and they all died. Once again Hitler's blunders were apparent.

In the next days, five German divisions were wiped out. By the end of the month, a continued drive westward shattered the German IX Corps which effectively destroyed the German Third Panzer Army. City after city and objective after objective of the Red Army were overrun. Anticipating well-prepared fixed defenses, each of the Russian attacking divisions was preceded by a company of T34 tanks fitted with mine-rollers, a heavy tank regiment, a heavy artillery regiment and an engineer assault battalion. Following these came a wave of flamethrower tank companies and light artillery regiments to eliminate pockets of resistance.

On June 26, the first of Hitler's reinforcements arrived from the Ukraine to plug the widening gap between the German Third Panzer and Fourth armies. With 70 Panther and 29 Tiger tanks, the 5th Panzer was sent to hold the German line east of the Berezina River until Busch's retreating Fourth Army could establish a proper defensive line. However, Busch's Fourth Army re-enacted a scene reminiscent of Napoleon's 1812 campaign—a rout. Masses of German troops were retreating from the east, abandoning their heavy equipment on the east side of the River, fleeing west in disorder, and crossing small crowded bridges under fire. Russian commander Rokossovsky exploited the collapse of German resistance in this sector, killing or capturing thousands of German soldiers.

As the Red Army drove across the Belorussian border, Hitler was slow to comprehend the gravity of Army Group Center's situation. On June 26, Busch and Ninth Army's General Jordan flew to Hitler's headquarters to try and persuade the Fuhrer to relent on his no-retreat policy. They complained it was destroying entire German armies a division at a time. Furious with the near-collapse of the Ninth Army, Hitler relieved both Jordan and Busch, replacing the latter with Field Marshal Walter Model, commander of Army Group North-Ukraine and Hitler's top troubleshooter. Things would not fare any better under Model, Hitler's stooge.

By June 29th, the German Ninth army had effectively been destroyed by both the Red Army and the Russian Air Force. That week, Rokossovsky's armies had killed about 50,000 Germans, captured another 20,000 (including 3,600 wounded prisoners who would be murdered by their Soviet captors). They also destroyed some 3,000 artillery pieces and tanks. Rokossovsky continued his drive northwest toward Minsk, seeking to trap Model's retreating Fourth Army, along with any remnants of the Ninth Army that might have escaped. By the time Hitler permitted the Fourth Army to retreat west of the Berezina River, there was almost nothing left to save. By the end of the operation, it had lost 130,000 of its 165,000 men.

On July 3rd, the Fuhrer concluded that Minsk was lost and allowed forces there to retreat. The next morning, Russian tanks entered Minsk, closing off another large pocket and trapping another 15,000 isolated German soldiers. As food and ammunition ran low for the units marooned there, they broke up into smaller formations and fled, quickly becoming vulnerable to the unforgiving wrath of partisan bands. Only 900 of the 15,000 trapped soldiers managed to reach German lines. On July 8th Model's Fourth Army ceased to exist.

Further north, other units of the German Third Panzer Army became isolated as a result of the Russian's rapid attack on Minsk and were quickly eliminated. Meanwhile, the Soviet high command ordered its exhausted troops westward to cities along the Polish and Lithuanian borders despite diminishing supplies of gasoline and ammunition.

As Model retreated, he tried to form a new defensive line between Lithuania and Ukraine. He took the remnants of the Ninth Army, reinforced them as best he could, and re-designated the thin line as a part of the German Second Army. Model was exceedingly vulnerable. However, the Soviet tanks began to run out of fuel and ammunition. This allowed Model to provide his forces in East Prussia and Poland a respite.

The Russian juggernaut was not finished. On July 8, Model's thin line broke and Vilnius, the capital of Lithuania, was surrounded. Despite Hitler's orders to hold Vilnius "at all costs," on the night of July 12-13, three-thousand German troops of 15,000 trapped in the city broke out, leaving the rest to suffer death or captivity when the city was taken on July 13. The cities of Pinsk and Grodno, on the Polish border fell on July 16th. The Third German Panzer Army's line collapsed by the end of July, forcing Model's northern flank onto German-Prussian soil. As Operation Bagration wound down, in several places, the Red Army had crossed the Niemen River, the traditional border between that portion of Russia and Poland. Some Russian units had reached the Gulf of Riga on the Baltic Sea, isolating Hitler's Army Group North. By mid-August Model's tenure as commander had run its course. He was transferred to the Western Front for a brief term as supreme commander in that crumbling theater.

Operation Bagration cost Germany 450,000 men (including 31 generals), hundreds of tanks, and more than 1,300 artillery pieces. Of the German troops lost, 160,000 were taken prisoner, half of whom were murdered on the way to prison camps or died in Soviet gulags. The Russians took 57,000 German prisoners and shipped them to Moscow. There they were paraded before Muscovites on July 17th, in part to refute Hitler's claim of a "planned withdrawal" from Belorussia.

But the battle was not as one sided as it might seem. Before pausing for the winter, the Russians had driven westward as far as Warsaw, Poland. But the Red Army had lost 765,000 troops, of which 178,000 were killed, 541,000 wounded, and thousands more missing. However, the ultimate strategic strength of the Soviet Union was its almost endless supply of reserves. Notwithstanding its losses; on July 13th, the Red Army launched a followup campaign in northern Ukraine, called the Lwow-Sandomierz offensive, with more than one million men. The offensive destroyed Army Group North-Ukraine, from which Hitler had sent divisions to help stop the collapse of Army Group Center.

By August of 1944, the German Fourth Army and almost all of the Ninth and Third Panzer armies ceased to exist. Thirty one German divisions had been destroyed and nearly 30 more were crippled. The Red Army was approaching the Vistula River in central Poland and had reached the outskirts of Warsaw. The Russians thereupon settled down to rest, refit, regroup, and prepare for the final assault against Germany and Berlin in late winter and spring of 1945.

Operation Bagration was a colossal victory for the Red Army. By July 3rd, Soviet forces had recaptured Minsk, capital of Belorussia, a city which had been in German hands for three years. By the end of July the Red Army had pushed into Polish territory, and had taken Lwow, the major cultural center of eastern Poland. By mid-August, some Soviet units had pushed as far west as central Poland, not that far from the German border.

Many historians agree that Operation Bagration was Hitler's worst military defeat of the war. Although ignored or glossed over in the west, it determined the fate of Nazi Germany perhaps more than any other one campaign in World War II. While the Americans and British were fighting to get out of Normandy in June and July 1944, the Red Army was in the process of destroying the German Wehrmacht in the east. The German army never recovered from the materiel and manpower it lost to Operation Bagration. An indication of the completeness of the Soviet victory is that 31 of the 47 German divisional or corps commanders involved were killed or captured.

Post Script

In August 1944, Romania rebelled against German occupation, joined forces with Russia, and declared war on Germany. Russian forces thereupon invaded Romania. In September, they invaded Bulgaria. In October, Russian forces invaded Hungary. Hitler's Germany now faced hostile forces to the southeast, the east, and west as Allied forces drove through France.

To the northeast, German Army Group North would be pushed by the Red Army into a pocket called the Courland Cauldron on the Courland peninsula in East Prussia. From January 1945 to the end of the war, they were trapped there. However, as many as 350,000 troops were evacuated by sea and air back into Germany proper as the Russians pressed against them. In one tragic incident, a German evacuation vessel, the Wilhelm Gustloff Hansa, carrying over 10,000 German troops and civilians, was sunk by a Russian submarine in the Baltic Sea. All but 500 perished. The mighty German Army Group North which had besieged Leningrad, retreated back to Germany with its tail between it legs, a shadow of its original force.

Hitler, the genius that he was, brought all this down upon himself when he invaded Russia in the first place. To the contrary, he was increasingly becoming Hitler the Dummkopf.

Breakout in France

While the Red Army was breaking the back of the Wehrmacht in Russia during June and July 1944, another major campaign would begin on August 1. After Operation Cobra under American general Omar Bradley, a hole had literally been blasted through the German lines south of St. Lo (just south of Omaha Beach). A new army would now begin to pour through that gap and take the war to the Germans.

That Army was the American Third Army under the leadership of General George Patton. Patton had been withheld in England prior to and during the first two months of Operation Overlord, publicly presiding over the fictitious American First US Army Group. He was a central figure in the Allied Operation Fortitude, designed to deceive Hitler and the OKW (i.e., German high command) the real invasion would come at Calais, France. That plan succeeded brilliantly.

In reality, Patton had been assembling and training the new American Army unit to be designated the US Third Army. In the latter part of July, it was transported across the English Channel onto the beaches at Normandy and began preparing for its engagement against the Germans. He commanded ten divisions (nine infantry, one armored), totaling 231,000 men. At 12 noon, August 1, 1944, the Third Army under Patton was activated. Patton had been assisted with the breach in the German line south of St. Lo through the intense bombing of July 25th followed by the attack of Bradley's actual First Army. The way for Patton's coming success had been paved by those American forces who fought their way out of Normandy, finally cut through the bocage country, and then blasted the hole through the German line in Operation Cobra.

Patton turned the Third Army southward into France behind the German lines and went "looking for trouble." The trademark of George Patton was aggressive warfare and his basic philosophy was attack, attack, attack. He was a general of the offensive and detested being on the defensive. Patton's aggressive action more often than not kept the Germans off balance and reeling in retreat. He was a master of mobile warfare, brilliantly combing armor with infantry. The Germans were renowned for their Blitzkrieg tactics early in the war. However, Patton literally out blitzkrieged the German army. They hardly knew what hit them. He was the antithesis of the calculating, plodding British Montgomery who seldom sent his troops into action before he had methodically planned, prepared, and plotted before attacking. Not surprisingly, Patton disdained Montgomery. The feeling was mutual.

Patton went searching for the enemy. He would use armored reconnaissance units to range ahead of his main forces, looking for them. Once found, he used his armored units to spearhead attacks. His armored infantry, supported by tanks and selfpropelled artillery, would then attack in force. Every break in German lines was exploited by more tanks which prevented the Germans from adequately regrouping.

Patton fully developed the doctrine of tactical air support by having the US Ninth Air Force fighter-bombers fly close-cover above his advancing columns. This tactic came to be known as armored-column cover and involved groups of P-51s or P-47s over each column. These were coordinated by a forward air controllers positioned in Patton's lead tanks on the ground. His Third Army headquarters had more staff dedicated to tactical air support and planning air strikes than any other ground army in Europe.

Patton was initially ordered by Eisenhower to capture French ports on the Atlantic coast in Brittany. Therefore, Patton's first operation was to push south and west with several divisions to cut off the Germans in Brittany, hopefully opening additional ports on the coast to Allied shipping. His units reached the coast in less than two weeks.

As the Third Army thus began its illustrious campaigns in France, Patton wrote this poem:

> So let us do real fighting, boring in and gouging,
> biting.
> Let's take a chance now that we have the ball.
> Let's forget those fine firm bases in the dreary
> shell raked spaces,
> Let's shoot the works and win! Yes, win it all!

That pretty well summed up his style of fighting.

As Patton's forces poured southward out of Normandy and Hodges, now the commander of the US First Army, continued mopping up and clearing out German pockets of resistance around St. Lo, Hitler hatched another blunder. The American Army, including Patton, was driving southward offering a tempting target for Hitler. He therefore ordered what remained of his battered Seventh Army in Normandy to attack the American flank. From Hitler's limited perspective back in Germany, far removed from the battle; he thought he could drive through the American flank, splitting the Americans and maybe even reverse the entire Normandy situation. However, the Wehrmacht in Normandy in the summer of 1944 was not the German Army of 1940 and was hardly capable of carrying out such an attack. Moreover, with the skies above filled with Allied fighter-bombers, massive Allied ground superiority, coupled with Allied knowledge of the German counterattack from intelligence intercepts, the Germans did not have a chance.

Three weak and run down panzer divisions began the counterattack on August 6. Led by powerful Panther tanks, the Germans met initial success and advanced up to seven miles into the American flank. However, the attack quickly bogged down against fierce American resistance. The US 30th Division bore the brunt of the attack but held out tenaciously during six days of fierce fighting, holding their position atop what was called hill 317. They lost almost 2,000 men, but stopped the German advance. The German counterattack accordingly failed.

Meanwhile, Patton's Third Army now swung around and began charging eastward. Hodge's First US Army took control to the north and the German Seventh Army suddenly found themselves in a precarious position. They were basically surrounded with Hodges and Montgomery above them to the north and Patton below them to the south. Their only hope was to escape to the east through a window near the French town of Falaise. On August 8, Montgomery ordered British and Canadian forces to begin driving southward to cut off the anticipated German retreat through Falaise. He, however, would be later criticized by the Americans for lacking urgency in actually getting his troops moving southward.

Hitler, finally realizing what was happening, ordered his Seventh Army as well as his Army Group B to retreat to the east through the open gap at Falaise. Patton also realizing what was happening turned northward toward Falaise, planning to entrap the Germans. Unbelievably, General Bradley Omar, now the overall American ground commander, ordered Patton to halt, utterly exasperating Patton. Bradley would later make the lame excuse that he feared the converging Canadian forces and Patton would collide with unhappy results. However, all Allied units had radios and the two converging forces certainly could have been forewarned and coordinated from attacking each other. The Russians and the Germans, with less robust communications, had sprung such traps numerous times on the eastern front. But Bradley prevailed, allowing tens of thousands of German troops to escape to the east. Nevertheless, the Germans were decimated within the Falaise pocket.

By August 17th, the encirclement of what was left of the German Seventh Army was complete. By August 22nd, all German forces west of the now closed Falaise gap were dead or prisoners of war. Historians have differed in their estimates of German losses in the Falaise pocket. However, the majority think from 80,000 to 100,000 troops were caught in the encirclement, of which 10,000 to 15,000 were killed. Forty to fifty thousand were taken prisoner and 20,000-50,000 escaped. The latter would soon be reorganized and rearmed in time to slow the Allied advance into the Netherlands and Germany. Though several German units had managed to escape on foot to the east, they left behind virtually all their tanks and heavy equipment. Bradley's timidity and Montgomery's unhurriedness prevented what became a partial victory from being a total triumph. The Normandy campaign was thus finished. Hitler, in sending his troops to attack the American flank, blundered once again and showed himself to be a dummkopf.

George Patton swore like a soldier and tended to be a prima donna. Accordingly, he led from the front in flamboyant style with a polished helmet and ivory-handled Colt 45 pistols, often standing up in his command car, siren blaring, as he drove amongst his troops. He pushed them hard, but his men loved him. His flamboyance aside, Patton was a well-trained tactician with an analytical military mind. He sought to avoid casualties by encircling German armies which he often did. He believed rambunctious American GIs fought best, when moving forward, especially in the summer. Then the weather tended to allow overwhelming American tactical air support. His soldiers could camp outside, while his speeding tanks had dry roads.

With the Germans retreating from the area, Patton ordered his Third Army in hot pursuit. Depleted German units were easily overcome. On August 25, 1944, lead elements of Patton's Army, arrived on the outskirts of Paris. Patton allowed the French Second Armored Division, under French General Philippe LeClerc, to take the lead in liberating their capital. Patton's division then marched through the heart of Paris in parade formation.

It took Patton only 30 days to sweep across France closing in on the German border. In early September 1944, as his forces neared the French province of Lorraine, the Third Army out ran its fuel supplies and ground to a halt, not far from the German border. Unfortunately, the Americans had struggled to supply their troops, especially Patton's army. But he finally ran out of gas, literally. He assumed he would be given priority of resupply due to the success of his offensive. He was frustrated to learn that would not be the case.

Red Ball Express

Though the Germans had pioneered the concept of Blitzkrieg with motorized infantry and fast tanks, they ironically still supplied their armies mostly with horsepower—literally horse-drawn wagons. The Americans had a completely different idea of logistics and troop supply. They used motorized transport and there were a number of reasons for this. America already was the number one automotive producer in the world prior to World War II. The long distances across America meant there already were regional trucking companies in operation when the war broke out. There thus was considerably more experience among Americans with trucks and motorized logistics than there was in Germany.

Once the US Army broke out of Normandy in the summer of 1944, American "army trucks" began carrying supplies to the front. These "army trucks" were mostly 2 1/2 ton, all-wheel drive "straight" trucks, though semis were involved as well. These were produced by American companies such as Studebaker, General Motors, Ford, White, and Dodge. But they pretty much all looked the same because they were built to the same Army specifications. As supplies came off the docks at Cherbourg or LSTs at Normandy, they were loaded onto these trucks. They then were organized in convoys and sent out to catch up with the frontline forces wherever they were.

This came to known as the Red Ball Express. American soldiers, mostly African Americans, drove these convoys around the clock, working in harsh conditions. Drivers did not get much sleep, but worked endlessly to get supplies to the front-line troops—especially gasoline. The express ran from August 25, 1944 until November 16, 1944. On August 29th the Red Ball

Express reached its highest tonnage day with nearly 6,000 vehicles carrying over 12,000 tons of supplies to the fight.

There was a shortage of trucks and drivers. Therefore, the Army commandeered trucks anywhere it could. Drivers for the Red Ball were not chosen for their driving skills, but for their availability. Soldiers who knew about as much about driving a truck as to how to do brain surgery were chosen, given brief training, "qualified," and put on the road. The established rules of the road were routinely ignored. Even though regulations were a speed limit of 35 miles per hour, many removed the governors from the trucks that allowed them to go 55 miles per hour. Some were even careening along the roads at up to 70 miles per hour. There were constant breakdowns and accidents as well as some sabotage by German prisoners used for routine tasks such as fueling and checking tire pressure. By the time the Red Ball Express was discontinued, the truckers had brought over 412,000 tons of supplies to the front, ensuring Allied success as it pushed into Germany.

And so, notwithstanding the Red Ball Express, George Patton's army was still out of gas at the end of September 1944. Eisenhower had ordered supplies be redirected eastward to the typically cautious Montgomery for his upcoming and ill-fated Market Garden operation. Meanwhile, the cutoff of Patton's supplies would prove disastrous. The scattered and retreating German forces were able to regroup, rest, and prepare for the next phase of the war. Their resistance stiffened as the weather grew worse and as the shortened German supply lines began to favor the Wehrmacht.

Operation Dragoon

As early as the Allied conference of the Big Three (Roosevelt, Churchill, and Stalin) at Tehran in December 1943, there had been plans to also invade France from the Mediterranean. Originally, the plan was for an Allied invasion to take place there at the same time as Operation Overlord in Normandy. However, a shortage of Allied shipping in general and landing craft in particular caused the operation to be cancelled. After the successful landings at Normandy, there were again available transport ships and landing craft. Over the strenuous opposition of Churchill, the Americans and Free French reinstated the plan. Churchill wanted an Allied invasion into the Balkans to block what he foresaw as Soviet dominance in that region after the war. He would prove to be correct in that prescience, but Roosevelt and Eisenhower prevailed. The plan to invade southern France was therefore reinstated to take place in mid-August 1944. The plan originally had been called Operation Anvil, but for security reasons, after being suspended for some months, was renamed Operation Dragoon.

The US Army drew three battle-tested divisions from the Italian front for the invasion and designated them as part of the US Seventh Army. The Free French likewise diverted three battle-experienced divisions from Italy and were designated as French Army B.

The objective of the operation was to secure vital ports on the French Mediterranean coast for the supply of American forces in France. It also was to increase pressure on the Wehrmacht in northern France by opening another front. After some preliminary commando operations, US forces landed on the beaches of the French Riviera August 15, 1944, between Toulon and Cannes. These were followed by several divisions of the Free French. They were opposed by the weakened German Army Group G under General Blaskowitz. Over the summer of 1944, the overall German commander of France, Field Marshal von Kluge, had steadily withdrawn divisions from Blaskowitz to help the German forces in desperate straits in Normandy. As a result, second and third-rate German soldiers with obsolete equipment opposed the Allied landing along the Mediterranean coast.

The landings went well for the Allies. On Dragoon's D-Day, August 15th, the Allies landed 94,000 men and suffered just 395 casualties. The First French Army quickly surrounded Marseilles and Toulon. Both cities fell to the Free French on August 28th, a full month earlier than anticipated.

Hampered by American air supremacy and a largescale uprising by the French Resistance, the weak German forces were swiftly defeated. The Germans retreated to the north through the Rhone valley to establish a defense line at Dijon. In the ensuing battle, the Germans, badly damaged and in full retreat, headed pall mall toward the German border. The operation formally ended in mid-September after the US Seventh Army hooked up with Patton's Third Army advancing from the west. The operation was a resounding success. It opened up new supply lines for Eisenhower, expelled the Germans from southern France, and provided the French with an opportunity to participate in their own liberation.

CHAPTER 30

Operation Market Garden

As the Americans broke out of Normandy and Patton raced across France, forces under the command of British General Bernard Montgomery began to move northeastward, more or less following the coast of the North Sea. With Patton knocking on the door of Germany and hopes that Montgomery could move eastward into Germany, there was talk that the war might be over by the end of the year. In the months following D-Day in Normandy, German troops began falling back wholesale as the allies forces advanced across northern France, Belgium, Luxembourg, and the Netherlands. By September 1944, however, the overstretched Allies were running into formidable defenses along the German Siegfried Line—a series of fortifications along the German border, similar to the Maginot line in France.

Montgomery came up with a bold plan to bypass the Siegfried Line by crossing the northern part of the Rhine River, then liberating and driving into the industrial heartland of northern Germany. The codename of the operation was called Market Garden. The plan called for three Allied airborne divisions (the "Market" part of the operation) to drop by parachute and glider into the Netherlands, seizing key territory and bridges. This in turn would allow Allied ground forces (the "Garden" part) to cross the Rhine. The attack was to be the largest airborne operation in World War II.

Eisenhower's greater objective was to encircle the Ruhr area, the heart of Germany's industrial region, in a giant pincer movement. The northern end of the pincer would have Montgomery bypass the north end of the Siegfried Line, allowing access into the north German plains, ideal for mobile warfare. Other American forces were expected to cross the Rhine further south, forming the southern part of the pincer. The goal of Operation Market Garden was to establish the northern end of that pincer and be ready to drive deeper into Germany. Allied forces would launch from Belgium, push 60 miles through the Netherlands, cross the Rhine, and consolidate north of Arnhem on the Dutch-German border—ready to close the pincer. The great hope was that this overall strategy would end the war by Christmas 1944.

Market Garden was a colossal failure. In fact, it was one of the few major defeats for the British and Americans after the early war years. The Americans did eventually cross the Rhine, but not until March of 1945. Operation Market Garden itself was completely unsuccessful.

On the morning of September 17, 1944, three divisions of the First Allied Airborne Army—the US 101st and 82nd Airborne and the British First Airborne—flew from bases in England across the North Sea to the Netherlands. The 101st Airborne's objective was capturing the Dutch city of Eindhoven, as well as several bridges over the canals and rivers north of that town. The 82nd Airborne was ordered to capture territory around the Dutch city of Nijmegen, including a key bridge over the River Waal. The operation required the seizure of the bridges across the Meuse River, two arms of the Rhine (the Waal River and the Lower

Rhine), along with crossings over several other smaller canals and tributaries. This was all in preparation for ground forces to follow. Allied intelligence had reports of a Panzer division in the area resting and refitting after the battles in Normandy. Unfortunately, operational planners ignored these reports.

The operation failed largely because the British 30th Infantry Corp failed to reach the furthest bridge at Arnhem before German forces overwhelmed them. (A famous movie was thereafter made about the attack called "A Bridge too Far.") Around 10,000 men of the First British Airborne Division landed near Arnhem. But their landing zones were seven miles from the bridge at Arnhem. Only one battalion reached the bridge while the remaining troops were forced into a small pocket by the Germans. Apart from a few antitank guns and artillery modified to fit inside their gliders, the lightly armed British airborne troops had no heavy weapons to resist powerful Panzer tanks. Although units of the British 30th Corp in conjunction with the US 82nd Airborne Division took the Nijmegen Bridge, they could not reach the farthest bridge at Arnhem. Most of their advance required passage along a narrow causeway, highly vulnerable to traffic jams from German counterattacks.

British field radios did not operate on the same frequency as RAF aircraft causing communication problems which reduced chances of success. Thick fog in England and low clouds over the battle zone also hampered both airborne resupply, airlifting of reinforcements, and tactical ground air support. The crossing of the Rhine and the capture of Germany's industrial heartland were delayed for six months. There would be no quick victory. Despite heroic efforts, the US-British forces failed to achieve their objectives and sustained devastating losses.

Though Operation Market Garden liberated much of Holland from Nazi control, it also established a foothold from which the Allies could make future attacks into Germany. The courage and determination of the Allied forces in Arnhem were evident, but Operation Market Garden was a costly failure. Of 10,600 Allied forces who crossed the Rhine in September 1944, 7,900 were killed, wounded, or taken prisoner. Overall Allied casualties during the operation was over 17,000, compared with around 8,000 German casualties.

CHAPTER 31

The Battle of Huertgen Forest

With Patton's Third Army pressing the attack against Germany along its southwestern border with France and Montgomery's failed Operation Market Garden to the north, American commanders attempted another disastrous attack in the north-central portion of the Allied front against Germany. It came to be known as the Battle of the Huertgen Forest.

By mid-September 1944, the Allied pursuit of the German army was slowing because of over-extended supply lines and increasing German resistance. The overall strategic goal was to move up to the Rhine River along its entire length and then prepare to cross into the heartland of Germany. Courtney Hodges, commander of the US First Army perceived a potential threat from enemy forces using the Huertgen Forest as a base. The Huertgen consisted of dense conifer forests and rugged terrain. In the fall and winter, heavy rain and snowfall, along with a lack of roads made it extremely difficult to penetrate, either by foot or in vehicles. Therefore vehicular movement was restricted. Conditions on the ground became a muddy morass, further impeding heavy vehicles such as tanks. Nevertheless, American commanders ordered their troops into its rough and unfamiliar terrain in order to continue their advance towards the Rhine in that sector.

The battle proceeded from mid-September 1944 to mid-February 1945, and ended with a weak American victory. However, the cost in human life was horrendous on both sides. Some historians have suggested that Hodges and his commander Omar Bradley were criminally derelict in driving American forces into the meat grinder of the Huertgen Forest. It was the longest battle fought by American forces in World War II and accomplished little tactically or in the greater strategy of the European Theater of Operations. A former company commander, who served in the Huertgen battle, described it as a misconceived and basically fruitless battle that should have been avoided."

The Americans had a distinct advantage in numbers, as high as five to one. However, armor, mobility, and air support were greatly reduced by weather, the forest, and terrain. A relatively small number of determined and prepared Germans added up to effective killing zones. To make matters worse, as American divisions suffered casualties, they were replaced by inexperienced troops, fresh from America, and sent into combat. More often than not, the results were disastrous.

The Germans faced the same problems of the terrain, forest, and weather. However, their commanders and many of their soldiers were veterans and battle hardened. They therefore were effective in fighting during the winter. Unfortunately, most of the green Americans were not much more than recent draftees rushed into the battle, not well-trained and inexperienced.

The battle of the Huertgen Forest ended in a dismal American victory. Though they technically prevailed in the end, the whole offensive was a virtual failure for the US. The Americans initially suffered 33,000 casualties and ranged upward of 55,000 casualties by the time it was all over, with a 25 percent casualty rate. The Germans also suffered heavy losses with 28,000 casualties. The Huertgen Forest campaign was one of the bloodiest battles in World War II for the American Army. Many historians in hindsight consider it to have been unnecessary, suggesting Allied forces could have simply bypassed Huertgen, leaving the German troops there stranded to whither on the vine while the Americans pressed onward around it.

CHAPTER 32

Patton at Metz

Still further south, George Patton's Third Army had run out of fuel in late September 1944. Following the American breakout from Normandy, the US Third Army had raced 400 miles across France, with Nazi forces retreating in disarray. As its supply lines lengthened, gasoline became scarce, Eisenhower ordered Patton to halt to conserve supplies intended for Operation Market Garden. This pause by the Third Army gave the Germans time to reorganize and fortify an area of the German West Wall around Metz with the hope of stopping Patton. Though with the pause of Allied forces on the western front, Patton was allowed to initiate limited attacks against German positions in early October.

The region of Metz, Germany, had not been considered of strategic value by the Germans until Patton had begun his race across France. Originally the Metz area was actually a complex of old fortresses built by the Germans dating back into the 19th century. It was a sprawling fortress system with subterranean forts spread to a depth of ten miles. The massive system, which made Metz the most heavily fortified city in Europe at the time, consisted of 43 forts arrayed in an inner and outer belt that together mounted 128 heavy guns. These wreaked havoc on Patton's forward positions.

The forts in the outer belt were located in close proximity to each other, enabling them to provide interlocking fire support. The two largest forts on the west bank of the Moselle River were named Driant and Joan of Arc. They contained revolving steel turrets housing their big guns. Their crews and the rest of each garrison were protected in subterranean barracks which were surrounded by dry moats and multiple rows of barbed wire. Direct assault against these forts would be almost suicidal, as Patton soon learned. Many of these forts were interconnected by underground tunnels so that troops, munitions, and supplies could be shuttled back and forth as needed.

Fort Driant was the strongest and most modern fort in the outer defensive ring surrounding Metz. It was five miles southwest of the city on the west bank of the Moselle River atop a high hill. This in turn was surrounded by rows of barbed wire on the outer perimeter. Then the close-in defense was a dry moat 60 feet wide and 30 feet deep intended to obstruct infantry and tracked vehicles. The fort itself was protected by a 15-foot-thick roof of reinforced concrete. And this was just one of the 43 forts in the system.

Moreover, as per orders issued by Hitler in March 1944, fortress commanders were directed to hold their positions at all costs, surrendering only with Hitler's approval—which he never gave. Thus the German commanders of the Metz fort system were under orders from Hitler to hold them at all costs when attacked in late September and early October 1944. Hitler understood that Patton's pause was due to supply problems which soon would be resolved. He certainly recognized Patton's approach as a direct threat to the Saar region of Germany. He therefore ordered commanders in that region to hold out as long as possible to give time for the strengthening of other border defenses which had been neglected to build up the Atlantic Wall. Thus the scene was set for a massive battle.

On the morning of September 27, P-47 fighter-bombers swooped low over the fort and dropped 1,000-pound bombs and napalm canisters at the base of the fort and on the structure itself. That afternoon, infantry with the support of tank destroyers, attacked from two directions but made no headway against enemy pill-boxes leading up to the fort. The following day no additional progress was made, and the attack was broken off.

The battle would rage for almost three months. Patton ordered attack after attack against the forts with minimal success. Various tactics were involved from bombing to artillery to flame-throwers to direct infantry assaults. Little progress was made. However, by mid-November, US forces had managed to isolate most of the forts, and began attacking the city of Metz proper. The Wehrmacht in Metz began retreating on November 17th, with American units pursuing them. The US Third Army entered Metz on November 18th. Although the city itself was captured by Patton and hostilities formally ceased on November 22nd, the remaining isolated forts continued to hold out into December with Fort Joan of Arc surrendering on December 13th.

The consequences of the American victory were several. The Germans had succeeded in delaying Patton from invading Germany proper, enabling them to make an organized withdrawal to the Saar River and to reorganize their defenses there. To that degree, they achieved a sort of victory. But the Americans had broken through the German West Wall. Perhaps less visible, the hubris and swagger of Patton and his Third Army had been tarnished. It had taken them only a month to drive the Germans out of their sector of France, but almost three months to overcome the obstacles at Metz. The battles had been bloody yet Patton and the Third Army prevailed. Winter had arrived and unbeknownst to Patton another major battle was looming which would divert his attention away from Metz and the invasion of Germany. That was the upcoming Battle of the Bulge. Meanwhile, Patton paused to allow his troops to rest, refit, and prepare for what lay ahead. Allied forces were closing in on Germany in a broad front from Montgomery in the Netherlands in the north to Patton on the German southwestern border.

CHAPTER 33

The Ardennes Offensive
(a.k.a., The Battle of the Bulge)

By late summer 1944, the handwriting was on the wall. The defeat of Germany was imminent. The American military knew it—they even hoped to wrap things up by Christmas of that year. The Russians knew it. Even high ranking German commanders knew it. Some had the temerity to discuss it openly, though when such talk reached the ears of Hitler, they were summarily relieved of command. After Operation Bagration, the Red Army was basically on the eastern border of the Germany. The Anglo-American armies were on the western border of Germany. The walls were closing in on Hitler. Any military officer with common sense knew the end of the war was nigh. The question was not if Hitler would be defeated, but only when.

Unfortunately, Adolf Hitler did not see things that way. He clung to a desperate hope that fate would look kindly upon him and through some miraculous intervention, Germany would turn the tables and win the war. Thus, the Fuhrer began planning a final offensive operation which would prove to be another major strategic blunder. When it was over, the Wehrmacht would have gained nothing, but lost a substantial portion of its military strength and resources. The end result was the Allies eventually had it that much easier in finishing off the Reich.

In late summer 1944, Hitler began privately planning a major offensive operation against the Americans and British. Initially, he told no one, not even his top military commanders. His immediate objective was to send a spearhead of tanks and armor through the Allied front to break out all the way to Antwerp, thus denying the Americans and British of its port facilities which were about to re-open. However, his greater strategic objective was to physically drive a split between the Americans to the south and British forces to the north. Success would also give the Germans time to design and produce more advanced weapons such as jet aircraft, new U-boat designs, and super tanks. They hopefully could also be able to then concentrate additional forces in the east.

Even more so, Hitler thought in so doing, he could break apart the Anglo-American alliance and sue for a separate peace, apart from the Russians. Hitler was aware of the friction between Montgomery, the top British commander, and top American commanders such as Bradley and Patton. He hoped to exploit that fissure. Many high-ranking German commanders also entertained the notion that perhaps they could persuade the Americans to join with them in the fight against the Soviets. Whether Hitler himself entertained such an idea is not clear. But for sure, he hoped to destroy the western alliance and in the process even encircle four Allied armies. Hitler had big ideas, but the chances of this grandiose plan actually coming to pass were next to nil, and his top commanders knew it when he revealed it to them.

On September 16, 1944, Hitler first officially outlined his surprise counteroffensive to his astonished top generals. They were sworn to secrecy. In fact, the actual operational commanders, Model and von Rundstedt, were not informed of their role until even later. Hitler knew the Allied front, the American line in particular, was thin in the Ardennes region of Belgium. Therefore, he proposed to his top commanders attacking through the American lines there. The initial plan was for German armor to pierce the thinly held lines of the US First Army by the end of the first day, get the armor through the Ardennes by the end of the seconD-Day, reach the Meuse River by the thirD-Day, and seize the Port of Antwerp and the western bank of the Scheldt estuary by the fourth day. That schedule would soon be significantly modified, but it would still be ambitious. Part of Hitler's rationale was that German forces had marched through the Ardennes in three days in 1940. But this time, he was up against far more powerful forces in the American army.

Hitler's counter-offensive came to have several names. The Germans called it "Operation Watch on the Rhine." They also called it the "Ardennes Offensive." However, as the battle would eventually unfold, the American press, noticing on maps the bulge being driven into the Allied line, began to call it the "Battle of the Bulge." And indeed, as the Germans would soon drive into the American line, they pushed it back creating a bulge or a pronounced salient.

With Patton's Third Army stalled in late September 1944, German commanders were able to reorganize and refit the troops driven back into Germany. This would come back to haunt the Allies when the Ardennes Offensive got under way. After the meat grinder of the battles of the Huertgen Forest, several American divisions were withdrawn to rest, recuperate, and refit in the Ardennes region. Moreover, after Patton's rout of the Wehrmacht across France and other advances by the American First Army further north, a spirit of complacency had descended particularly upon American troops. Most of the Americans thought the war would be over by Christmas. Even top American commanders considered the German army beaten and no longer able to mount a major offensive. Though still a force to be reckoned with, Allied forces from the top down considered Germany all but defeated. They were in for a rude awakening.

The Plan

Hitler's plan was extensive, daring, and complex. Knowing full well American air supremacy over the battlefield, Hitler purposefully planned his counter-offensive for early winter when skies were typically overcast and flying weather was poor. The attack was initially planned to begin on November 27 but was delayed until December 16 due to problems accumulating necessary fuel for the operation. Hitler and his staff concluded that for the operation to succeed, four criteria had to be met. (1) The element of surprise was crucial; (2) the weather must be poor to suppress Allied air operations; (3) the progress had to be rapid— the Meuse River, halfway to Antwerp, had to be reached quickly; and (4) Allied fuel supplies would have to be captured along the way because Germany was out of fuel. The German General Staff estimated they only had enough fuel to cover one third to one half of the ground to Antwerp in heavy combat conditions.

Hitler never had much confidence in his generals. But after the assassination attempt against him on July 20th by Wehrmacht officers, he no longer trusted any of them. Therefore, he assigned the key leadership roles of the Ardennes Offensive to trusted Waffen-SS officers. The latter were members of the Nazi Party whose loyalty to Hitler was unquestioned. The lead role in the attack was thus given to the Sixth Panzer Army, commanded by SS Colonel Sepp Dietrich.

In typical Hitlerian strategy, three battle groups or armies were formed. Army Group North, the northernmost battle group and closest to the final objective of Antwerp, was commanded by Sepp Dietrich. It included the most experienced formation of the Waffen-SS: the First SS Panzer Division Leibstandarte Adolf Hitler, along with the 12th SS Panzer Division Hitlerjugend. The latter were comprised of older teenagers of the Hitler Youth organization who had been indoctrinated throughout their formative years in absolute loyalty to Hitler. They were renowned as being fanatical in their fighting. These received the first priority of supplies and equipment. Army Group Center, commanded by General von Manteuffel, had as its objective the capture of Brussels. Finally, Army Group South was commanded by General Brandenberger with the task of protecting the southern flank of Manteuffel's group.

Hitler scraped together 30 mostly under-strength divisions for his three battle groups. Among them were Volksgrenadier (People's Grenadier) units formed of veterans combined with recruits formerly regarded as too young, too old, or too frail to fight. Training time, equipment, and supplies were inadequate for these prior to the coming offensive.

When the Wehrmacht occupied France, the French Resistance supplied the Allies with a constant stream of intelligence on what the Germans were up to. Now on their home soil, that intelligence had dried up. Because secrecy was crucial for a surprise attack, communications between German units were accomplished by land-line telephones and teletype machines. However, having learned the hard way of both Allied and Russian deception campaigns, Hitler now employed a little of his own. He knew he could not keep such a massive troop buildup along his border completely secret. Therefore, German forces sought to convince the Allies that troop movements to the border areas were for defensive purposes. The Wehrmacht purposefully used radio traffic about strengthening their defenses to convince

Allied intelligence that troops accumulating along the German border were there to defend Germany against Allied attack. It worked. And, because of foul weather in late November and early December, Allied air reconnaissance was greatly restricted. American commanders really did not know what was going on.

Much of the troop deployment and supply convoys toward the border was done at night. German troop formations were forbidden to make unnecessary noise. When heavy truck traffic was at hand, German artillery would fire enough harassing rounds that the noise thereof covered the rumble of the trucks. Soldiers in the field were forbidden to have open campfires at night on which to cook. They were restricted to using small amounts of charcoal for heat and cooking. Thus, the first criteria of Hitler's plans was effective. Most of the Americans did not have a clue as to what was building up across the front line.

Nevertheless, Patton's Third Army intelligence chief along with Hodge's First Army intelligence chief both accurately predicted the German offensive and intention to strike. These warnings were sent up through the chain of command to Omar Bradley, the commander of the American 12th Army Group. Ironically, Bradley's response was "Let them come." He basically dismissed the reports.

Because the Ardennes sector of the American front line was considered quiet, the US high command directed it to be used as a training ground for new units arriving from America and as a rest area for units that had seen hard fighting. The US units deployed to the Ardennes sector were therefore a mixture of inexperienced troops and battle-hardened veterans sent there to recuperate. The US 99th division had just arrived from America along with the 106th Division. The American 28th Infantry Division was also in the area recuperating from the vicious fighting in the

Huertgen Forest. Ironically, it would be the 99th and the 28th divisions which would play a major role in turning back the German offensive.

Hitler also had more devious schemes up his sleeve. Along with the direct military campaign, he with his staff, concocted what they called *Operation Greif*. Hitler had high regards for an SS commando by the name of Otto Scorenzy. He was the man who had engineered the daring rescue of Benito Mussolini after he had been deposed and arrested. Scorenzy was a master of special operations and dirty tricks. He along with Hitler devised a scheme in which he would lead a task force of English-speaking German soldiers, complete with American "accents" in Operation Greif. These soldiers were to be dressed in captured American uniforms, wearing dog tags taken from dead American soldiers and prisoners of war. Their mission was to go behind American lines and change signposts, misdirect traffic, generally cause disruption, and seize bridges across the Meuse River. In late November another ambitious project was added. A paratrooper combat group in what was called "Operation Stoesser" would do a night drop behind the Allied lines aimed at capturing a vital road junction near Malmedy, Brussels. As it turned out, that operation failed.

Hitler thus had put together a grandiose scheme which he thought could give victory to Germany and end the war in the west. But only two of his crucial criteria came to pass. The rest did not. The Germans attack was a complete surprise upon the Americans. They were caught completely off guard. Second, the weather was lousy, at least for the first week of Operation Watch on the Rhine. The other two crucial criteria never materialized. Though crashing through unsuspecting American lines and racing toward their goals, the advance quickly bogged down and eventually ground to a halt. Though German forces were able to capture some American gasoline, they never were able to seize enough to keep their armies going.

The Attack

On December 16, 1944, at 5:30 in the morning, the German Army Group North began the attack with a powerful 90-minute artillery barrage. Steel and fire from 1,600 artillery guns rained down on American troops across an eighty-mile front facing Sepp Dietrich's Sixth Panzer Army. The American's initial reaction was that this was an anticipated localized counterattack. (The Allies had recently attacked a German sector to the north and the Second Division had knocked a sizable hole in the Siegfried Line.) Meanwhile, heavy snow had fallen on parts of the Ardennes region that day. While keeping Allied aircraft grounded, the weather also was a problem for the Germans. Poor road conditions slowed their advance which led to massive traffic jams. With many tank and truck engines idling, this only exacerbated fuel shortages in forward units.

In the central sector, von Manteuffel's Fifth Panzer Army attacked towards Bastogne and St. Vith. These were both major road junctions of strategic importance. In the southern sector, Brandenberger's Seventh German Army pushed towards Luxembourg in its efforts to secure the southern flank from Allied attacks.

Battle of Elsenborn Ridge

The Siege of Bastogne, at least in popular American history, is considered to be the focal point of the Battle of the Bulge. However, the battle for Elsenborn Ridge would become the decisive struggle of the Ardennes Offensive. There, American troops blocked the advance of the best equipped Panzer armored units of the German army. There, experienced veterans alongside of green Americans units forced cocky SS troops and their Hitler-youth counterparts to seek unfavorable alternative routes for their objectives, considerably slowing their advance.

The units attacking Elsenborn Ridge had been personally selected by Adolf Hitler. Dietrich's Sixth Panzer Army was not only given priority for supplies and equipment but was given the shortest route to the ultimate objective of the offensive, Antwerp. The Sixth Panzer Army included the elite of the Waffen SS, which fielded four Panzer divisions and five infantry divisions.

SS Lieutenant-Colonel Joachim Piper, of the Sixth Panzer group, led a brigade consisting of 4,800 men which was the vanguard of the German attack. Piper's group included Germany's new and most powerful tank, the Tiger II (sometimes called the King Tiger). Head-to-head, it could obliterate any American tank. However, the Tiger II consumed 2 US gallons of fuel per mile (or, .5 miles per gallon). To make matters worse, the Germans had only about one third of the fuel they needed to reach Antwerp. To address the lack of fuel, Piper had been provided with a map showing American fuel depots from which he could seize fuel if needed. However, with one exception, he failed in that regard. The attacks by the Sixth Panzer Army's elite infantry units faltered when they ran into the US Second and 99th Infantry Divisions, the latter was a green unit. US Army engineers blew up crucial bridges in the path of Piper's advance forcing him to find alternative routes.

In another instance, a platoon from the 99th Infantry Division (about 40 soldiers) along with four forward air controllers held up a battalion of about 500 German paratroopers until sunset, causing 92 casualties among the Germans. This further delayed

Piper's advance and prevented him from dominating roads to the south and east and bringing up needed fuel and supplies.

The Malmedy Massacre

On December 17th, shortly after noon, Piper encountered elements of an American Observation Battalion of the US Seventh Armored Division. After a brief battle the lightly armed Americans surrendered. They were disarmed along with approximately 150 other Americans captured earlier. They were then sent to stand in a field near a crossroads under light guard. About fifteen minutes later, a SS unit commanded by a Major Poetschke drove up. The SS men immediately opened up on the American prisoners with machine guns. As soon as the firing began, most were shot where they stood, but some managed to escape to tell their terrible story. Accounts vary, but at least 84 American prisoners were mowed down. Such actions were common by the SS on the Russian front, but now as Germany was in desperate straits, they did the same to American troops. Following the war, soldiers and officers of combat group Piper, including Joachim Piper and SS general Sepp Dietrich, were tried as war criminals for the incident at Malmedy. They were convicted and hanged. Sadly, the Malmedy massacre was not the only murder of American POWs by Piper. Other smaller groups were also gunned down by his SS men.

The Battle Continued

On December 19th, Piper's spearhead had pushed north to engage the inexperienced US 99th Infantry Division near the town of Stavelot. Piper was already far behind his timetable due to stiff American resistance. Furthermore, when the Americans fell back, US Army combat engineers blew up bridges and emptied fuel dumps. Piper's unit had counted on capturing that fuel and was further delayed, his vehicles again denied critically needed fuel. It took 36 hours for Piper to advance from the Eifel region to Stavelot. It took the German army nine hours to make the same advance in 1940. Meanwhile, American P-47 Thunderbolts were arriving and shooting up Piper's column causing further problems. As lead tanks were put out of action, the narrow roads were blocked until the burned out hulks could be removed. Battle after battle raged as German units ran into unexpectedly stiff American resistance. Moreover, American combat engineers continued to blow up bridges ahead of Piper further hampering his advance. Piper now faced being completely cut off by American units. Efforts to re-supply him failed when they were interdicted by American forces.

Finally, on December 23rd, one week after the German assault began, Piper ran out of fuel. Far from his objective, he was in danger of being surrounded. Without permission from Hitler, 800 of his men dismounted their tanks and vehicles and escaped on foot back to German lines. The German Army Group North had failed. They were the elite of the elite of the German army. They had utterly surprised recently-arrived American troops of the 99th Division. Yet, those green Americans stopped the mighty Waffen-SS Panzer unit.

The US 99th Infantry Division was outnumbered five to one but inflicted casualties of eighteen to one against the Germans. The 99th lost about 20% of its effective strength, including 465 killed and 2,524 men evacuated from injuries, fatigue, or trench foot. German losses, however, were much higher. In the northern sector opposing the 99th, Dietrich's vaunted Sixth Panzer Army suffered more than 4,000 deaths. Dwight Eisenhower's son, John, a military historian, would later write, "the action of the Second and 99th Divisions on the northern shoulder could be considered the most decisive of the Ardennes campaign." The stiff American defense prevented the Germans from reaching the vast supply and fuel dumps leading to the Meuse River. The crucial prong of Hitler's offensive had failed.

Ike's Conference

Meanwhile, on December 17th, one day after the attack began, Eisenhower realized what was happening. Contrary to initial conclusions by subordinates that the German attack was just a localized counterattack, Ike understood a full-fledged German counter-offensive was under way. He therefore called a war council of his top commanders of the European Theater of Operations. They met on December 18th in a cold bunker at Verdun, France. By this time, the town of Bastogne and its network of 11 hard-topped roads leading through the widely forested terrain of the Ardennes was under severe threat. By this time also, Hitler's Army Group Center under Manteuffel's Fifth Panzer division had penetrated deeply into their sector and were in the process of surrounding Bastogne. The only corridor open to and from Bastogne (to the southwest) was threatened and it had been sporadically closed as the front shifted. American leadership anticipated the road would be completely closed sooner rather than later, ensuring the town would be surrounded. American forces in Bastogne were in dire straits.

Ike told his generals, "The present situation is to be regarded as one of opportunity for us and not of disaster. There will be only cheerful faces at this table." Of course, Eisenhower realized that in extending their salient westward, the German Army became a sitting duck for powerful flank attacks against the base of their bulge (i.e., the salient). Ike also knew that once fully mobilized, the Allied forces in the region had twice the number of troops the Germans had, vastly more materiel, and importantly, an almost endless supply of fuel, not to mention air supremacy. Within a week of the attack, an additional 250,000 Allied troops would join the battle against the Germans. But at the moment, the crisis was at Bastogne.

Ike asked Patton how long it would take to turn around his Third Army to move north and counterattack the German flank. Elements of the Third Army were located about 100 miles to the south. Such a maneuver was a monumental task for any army. Under the present severe winter conditions, it was herculean. To the disbelief of the other generals present, Patton replied that he could attack with two divisions within 48 hours. Ike thought Patton was either being braggadocios or trying to be cute. He told him to be serious. But unknown to the other officers present, before he had left for the conference, Patton had ordered his staff to prepare three contingency plans for a northward turn for at least several divisions. By the time Eisenhower asked him how long it would take, that movement was already underway.

Bastogne

The Germans fared better in their central sector as Manteuffel's Fifth Panzer Army attacked positions held by the US 28th and 106th Infantry Divisions. Manteuffel did not have the overwhelming strength that Dietrich had in the north. However, he still had a superiority in troop strength and armor over the very thinly spread American defenders. As Manteuffel attacked the American positions, at least seven thousand men, probably more, were either casualties or taken prisoner. They also lost large amounts of weapons and materiel. The battle known as Schnee Eifel became the most serious defeat of the American Army in the final two years of the war in Europe. The town of St. Vith, another important road junction also fell, but in so doing, slowed the German time table. Unlike the German forces on the northern and southern shoulders of the bulge which experienced considerable difficulties, Manteuffel's advance in the central sector gained considerable ground.

Meanwhile, Otto Scorenzy was successfully infiltrating segments of his English-speaking Germans in American uniforms behind the Allied lines, particularly in the central sector of the battle. Although they failed to seize the crucial bridges over the Meuse River, their presence caused great confusion and rumors spread quickly. Even Patton was alarmed. On December 17th, one day after the initial German attack, he described the situation to Eisenhower as Krauts speaking perfect English creating havoc, cutting telephone wires, turning road signs around, spooking whole divisions, and shoving a bulge into American defenses.

Checkpoints were set up all over American rear areas, hindering crucial movement of soldiers and equipment. There, American MPs questioned troops on things every American should know—such as the identity of Mickey Mouse's girlfriend, baseball scores, or the capital of a particular American state. General Omar Bradley was briefly detained when he correctly named Springfield as the capital of Illinois but the American MP who questioned him thought its capital was Chicago.

The heightened security nevertheless made things difficult for the German infiltrators and numbers of them were captured. Even when interrogated, they continued their objective of spreading disinformation. When asked about their mission, some claimed they had been ordered to go to Paris to either kill or capture General Eisenhower. Security around the general was therefore greatly increased and Ike was confined to his headquarters. However, when Scorenzy's men were captured in American uniforms, they were executed as spies.

On December 21st the Germans accomplished surrounding Bastogne. The town's defenses were now in the hands of the recently arrived US 101st Airborne Division, the all African-American 969th Artillery Battalion, and elements of the 10th Armored Division. Conditions were difficult. Most medical supplies and medical personnel had been captured. Food was scarce. By December 22nd artillery ammunition was restricted to 10 rounds per gun per day. It was one of the hardest winters in Europe in many years. Soldiers huddled in frozen foxholes, forbidden to light cigarettes or fires at night. The German's shelled the town continuously. Anything that moved on the American front line drew sniper fire. The elements were as much of an enemy as were the Germans.

Patton's Prayer

One man was concerned about the miserable weather—George Patton. As the Battle of the Bulge erupted, Patton called for Chaplain O'Neill, Third Army Chaplain. The general informed his chaplain that he wanted a prayer to God be prepared regarding the miserable weather. The chaplain therefore sat down and typed out a prayer on a 3x5 card, noted below. After the war, Chaplain O'Neill wrote about the incident.

Patton sat me down and asked, "Chaplain, how much praying is being done in the Third Army?" "Does the general mean by chaplains, or by the men?" asked O'Neill. "By everybody," Patton said. The chaplain replied, "I a.m. afraid to admit it, but I do not believe that much praying is going on. When there is fighting, everyone prays, but now . . . men just sit and wait for things to happen. Prayer out here is difficult. Both chaplains and men are removed from a special building with a steeple. . . . I do not believe that much praying is being done."

"Chaplain," Patton continued, "I a.m. a strong believer in prayer. There are three ways that men get what they want; by planning, by working, and by praying. Any great military operation takes careful planning, or thinking. Then you must have well-trained troops to carry it out: that's working. But between the plan and the operation there is always an unknown. That unknown spells defeat or victory, success or failure. It is the reaction of the actors to the ordeal when it actually comes. Some people call that getting the breaks; I call it God. God has His part, or margin, in everything. That's where prayer comes in. Up to now, in the Third Army, God has been very good to us. We have never retreated; we have suffered no defeats, no famine, and no epidemics. This is because a lot of people back home are praying for us. We were lucky in Africa, in Sicily, and in Italy. Simply because people prayed. But we have to pray for ourselves, too. A good soldier is not made merely by making him think and work. There is something in every soldier that goes deeper than thinking or working—it's his 'guts.' It is something that he has built in there: it is a world of truth and power that is higher than himself. Great living is not all output of thought and work. A man has to have intake as well. I don't know what you call it, but I call it . . . prayer, or God."

The chaplain went on to say that Patton talked about Gideon in the Bible and said that men should pray no matter where they were, in church or out of it. And, that if they did not pray, sooner or later they would "crack up."

Thereafter, Patton called in an Army printer and had the field-topographical company print the chaplain's prayer on a small-sized card, making enough copies for distribution to the entire army. On the back, General Patton include a Christmas greeting to the troops. The card was made up, printed, and distributed to the troops on December 21st. The prayer is below.

The prayer read,

> "Almighty and most merciful Father, we humbly beseech Thee, of Thy great goodness, to restrain these immoderate rains with which we have had to contend. Grant us fair weather for Battle. Graciously hearken to us as soldiers who call upon Thee that, armed with Thy power, we may advance from victory to victory, and crush the oppression and wickedness of our enemies and establish Thy justice among men and nations."

Though Patton was known for his toughness and strong language when dealing with men, insight into his spiritual character was revealed by this conversation as recorded by the chaplain. Patton also kept a Bible on his night stand beside his bed and read therefrom each day.

The Weather Cleared

The weather, cleared the next day, December 22nd. Supplies, especially ammunition, were air-dropped over the next several days. Clearing weather also allowed American fighter-bombers to begin attacking German positions from the air. With the winter solstice at hand, the days were the shortest of the year, with darkness falling by 4:30 in the afternoon and the sun not rising until 8 a.m. in the morning. Yet the cold, clear days brought some encouragement by virtue of the sunshine. Nevertheless, the Germans continued to tighten the noose around Bastogne. The American soldiers were outnumbered approximately five to one and most of them lacked cold-weather gear.

On December 22nd the fortunes of the American forces in Bastogne seemed hopeless and the Germans knew it. The

German commander, General von Lüttwitz, sent a party under a white flag of truce to the American lines with a demand for surrender. The note was passed on to the American commander in Bastogne, Brigadier General Anthony McAuliffe

Lüttwtiz note read:

> To the U.S.A. Commander of the encircled town of Bastogne.
>
> "The fortune of war is changing. This time the U.S.A. forces in and near Bastogne have been encircled by strong German armored units. More German armored units have crossed the river Our near Ortheuville, have taken Marche and reached St. Hubert by passing through HompreSibretTillet. Libramont is in German hands.
>
> There is only one possibility to save the encircled U.S.A. troops from total annihilation: that is the honorable surrender of the encircled town. In order to think it over a term of two hours will be granted beginning with the presentation of this note.
>
> If this proposal should be rejected one German Artillery Corps and six heavy A. A. Battalions are ready to annihilate the U.S.A. troops in and near Bastogne. The order for firing will be given immediately after this two hours term.

All the serious civilian losses caused by this artillery fire would not correspond with the well-known American humanity.

The German Commander.

When McAuliffe, acting commander of the 101st, was informed of the Nazi demand to surrender, he simply said, "Nuts!" After turning to other pressing matters, his staff reminded him that he should respond to the German demand. One officer suggested that McAuliffe's initial reply would be "tough to beat." Therefore, McAuliffe had the word typed up and delivered to the Germans. That line became famous and a morale booster to his troops: "NUTS!"

When the one-word note was delivered, the officer leading the German truce party read it. Puzzled, he asked what it meant. The American officer replied, "It means go to h..., bud."

On December 23rd, the skies were again clear allowing the American Ninth Air Force to attack German positions. Devastating bombing raids were made on German supply depots in their rear. P-47 Thunderbolts attacked German troops on the roads. Meanwhile, much-needed supplies—medicine, food, blankets, and ammunition were dropped by parachute into Bastogne. A team of volunteer surgeons were flown in by military glider and began operating in a make-shift operating room located in an old tool shed.

The siege of Bastogne continued. On Christmas Day 1944, a large German Panzer unit attacked Bastogne in several places on the west side of the town rather than launching one simultaneous attack on all sides. The assault, despite initial success in penetrating the American line, was defeated and all the German tanks involved were destroyed. But the situation remained grim. One more German assault would likely succeed with Bastogne being overrun, the 101st destroyed or taken prisoner, and the way open for Hitler's attack toward Antwerp to succeed.

However, the next day, December 26th, tanks of Patton's Fourth Armored Division, along with his 26th Infantry Division, arrived from the south, broke through the German line, and opened a corridor to Bastogne. They had driven about 100 miles in four days over wintery roads. In the days following, more of Patton's army arrived strengthening the corridor already opened and relieving the siege of Bastogne. Manteuffel's Army group center was thwarted. For all intents and purposes the Battle of the Bulge had been won. Manteuffel recommended to Hitler a halt to all offensive operations and a withdrawal back to their starting point, the Westwall. Hitler, of course, rejected this.

Though his commanders on the ground knew the score, Hitler continue to flail away, fighting a losing battle. Bitter fighting continued in areas away from Bastogne well into January. Though the German offensive petered out in January 1945, they still controlled a dangerous salient in the Allied line. Eisenhower ordered Patton's Third Army to the south around Bastogne to attack northward. Montgomery's British forces to the north were to strike southward in a pincer movement and trap the remaining German units. The two Allied forces were to meet at Houffalize, Belgium. Montgomery, however, refused to risk what he thought was his under-prepared infantry in a snowstorm. He did not launch his portion of the attack until January 3rd, which allowed a substantial number of German troops to withdraw and successfully escape, though losing most of their heavy equipment. Once again, as at Falaise the previous year, Montgomery's lethargic response prevented a major Allied victory.

On January 7, 1945, Hitler finally allowed the ending all *offensive* operations in the Ardennes. On January 14th, Hitler gave von Rundstedt permission to begin a retreat altogether from the Ardennes region. The Houffalize and the Bastogne fronts were abandoned. However, it was a fighting retreat. Hostilities continued for another 3 weeks. St. Vith was recaptured by the Americans on January 23rd, with the last German units participating in the Ardennes offensive returning to where they started on January 25th.

Casualties for the First and Third US Armies were over 75,000 (8,400 killed, 46,000 wounded and 21,000 missing in action). The Battle of the Bulge was one of the bloodiest battles for US forces in World War II. The German High Command estimated they lost between 81,834 and 98,024 men in Operation Watch on the Rhine of which 12,652 were killed, 38,600 were wounded, and 30,582 were missing.

Although the Germans began their counter-offensive with complete surprise and enjoyed some initial success, they quickly bogged down. Something so prosaic as running out of gasoline caused the greater operation to run out of steam. While Hitler did not achieve any of his goals, the Battle of the Bulge inflicted heavy casualties on the Americans and delayed the Allied invasion of Germany by several weeks. For the Americans, the loss of men, tanks and other materiel were quickly replaced. For the Germans, however, their losses were irreplaceable. Germany no longer had any reserves of anything. The close to 100,000 men lost in the battle were 100,000 fewer men that Germany had to try and stop the coming Allied invasion of the fatherland from the west and the Russians from the East. German industry and the transportation system had been wrecked by years of Allied bombing. They could not replace the tanks, artillery, trucks and other military equipment lost in the Ardennes offensive. Even if they could build more, they could no longer transport them to where they were needed. Hitler's grandiose plan to win the war in the Ardennes operation was another of his colossal blunders. He not only lost the battle, he ensured the coming final defeat of the Wehrmacht would be that much easier. Operation Watch on the Rhine was a desperate blunder by a desperate despot which had no chance of succeeding. It marked the beginning of the end for the Third Reich.

Epilogue

The Battle of the Bulge against the Nazis was carried out almost entirely by the US Army. The British made minor contributions. They could have done more if Montgomery had been aggressive at the battle of Houffalize. But after the Ardennes was largely won and the dust had settled, Montgomery publicly left the impression that he and the British army had won the day. At a press conference on January 7, 1945, Montgomery, despite positive remarks about American soldiers, gave the overall impression that he and the British deserved the lion's share of credit for the success of the campaign. Moreover, he implied he had been responsible for rescuing the besieged Americans. Needless to say, this infuriated American commanders and troops. Eisenhower had tried throughout the war years to hold the Anglo-American alliance together. But Montgomery's self-serving statements did much damage to it. Though a war hero to the British population, many American officers already had come to dislike Montgomery. He was seen as a hesitant commander, who was arrogant, and willing to say unkind words about the Americans. Winston Churchill found it necessary in a speech to Parliament to explicitly state that the Battle of the Bulge was purely an American victory. He said, "This is undoubtedly the greatest American battle of the war and will, I believe, be regarded as an ever-famous American victory."

Montgomery later said, "Distorted or not, I think now that I should never have held that press conference. So great were the feelings against me on the part of the American generals that whatever I said was bound to be wrong. I should therefore have said nothing." Eisenhower later said in his memoirs: "I doubt if Montgomery ever came to realize how resentful some American commanders were."

Bradley and Patton both threatened to resign unless Montgomery's command was relieved. Eisenhower, encouraged even by his British deputy Arthur Tedder, had decided to fire Montgomery. However, intercession by Montgomery's and Eisenhower's Chiefs of Staff, persuaded Eisenhower to reconsider and allowed Montgomery to apologize. That he did, but the damage had been done. The Battle of the Bulge was over.

CHAPTER 34

The Beginning of the End

A fter the Battle of the Bulge, the western Allies spent the month of February 1945 fighting across the Rhineland to reach the banks of the Rhine River, a distance ranging from 15 to 80 miles depending on one's location. Though Germany had its "west wall"—the Siegfried Line, the Rhine River was the major natural defensive barrier to the heart of Germany. The various Allied armies—Montgomery's Second British and Canadian Armies, Hodges' First American Army, Patton's Third and the Seventh Army—all moved across the Rhineland during February to the banks of the Rhine River. In places there were high bluffs along the lower Rhine where the river is a quarter of a mile across. It is a major river and a natural defensive line. Since the days of the Roman Empire it has served as central Germany's traditional defense against invasion from the west. The US Army Corps of Engineers determined the river to be totally un-fordable, even at low water. Moreover, the Germans had either destroyed or were preparing to destroy every significant bridge. Now, over two million Allied troops were poised along its banks to cross into the heartland of Germany.

On the eastern front, the Russians had paused during the early winter months to rest and refit. They sealed off what was left of Hitler's Army Group North in the Baltic States. They then overran German East Prussia along with the northern region of former Poland. To the south, they invaded Hungary, conquering Budapest, and were moving toward Germany from the southeast. The Russians were thus poised for their major drive against Berlin. By the end of February 1945, six million Red Army soldiers were on the banks of the Vistula and Oder Rivers, the latter only 40 miles east of Berlin. Facing them were about two million German troops.

Crossing the Rhine

The western Allies were preparing to cross the Rhine for the final drive into the heart of Germany. In the north, Montgomery with his Second British and Canadian armies (along with several American divisions) was preparing a campaign to cross the lower Rhine. Montgomery was a general who believed in total preparation for a looming battle. He therefore spent the winter months of early 1945 assembling not only troops for the crossing, but innumerable amounts of equipment, including amphibious landing craft, bridging material, and endless requisite engineering gear to get the job done. The scope of his preparations paralleled that of the Normandy landings on D-Day 1944. His plan was called "Operation Plunder."

Smoke screens were to be laid down beginning on March 16 to obstruct the build up from German eyes. Parachute and glider landings were planned to drop paratroops across the river behind German front lines. Four thousand artillery guns would fire for four hours before the crossings. And, RAF bombers would bomb German defenses just prior to the assault. Engineering units were to prepare pontoon bridges during the night prior to the attack. It was a monumental, elephantine operation akin to the landings at Normandy. In fact, it was to be the "official" planned Allied crossing of the Rhine. Montgomery perhaps hoped it would atone for his bungled attempt at Arnhem six month's earlier.

Meanwhile, the US Army also recognized they would most likely have to make an amphibious crossing of the Rhine to drive into central Germany. The most advantageous areas were somewhere north of Bonn, where the river entered relatively open and therefore more tank-friendly terrain. American planners largely ignored the Remagen area, about fifteen miles south of Bonn because the terrain on the east side of the river was rough and difficult for tank operations. There, the colossal Ludendorff railroad bridge remained standing. It had not as yet been destroyed by German engineers.

The railroad bridge had been built, primarily by Russian prisoners of war, between 1916-1919 and spanned 1,200 feet. On the east end of the bridge, a railroad tunnel carried rail traffic eastward. It therefore seemed an unlikely objective for the Americans. German engineers had rigged the bridge with explosives, removing them for a time to avoid detonation during a recent Allied bombing raid. They then were replaced as the American army approached. German infantry units guarding the bridge were weak.

The American First Army commanded by General Courtney Hodges had been pursuing the German 15th Army eastward in early March 1945. On March 6th, remnants of the Fifteenth Army retreated across the Remagen Bridge as the Germans prepared to detonate explosives and demolish it before the Americans showed up. Meanwhile, men and vehicles of the US First Army approached the bridge, hoping but hardly expecting that they could seize it intact.

German troops kept retreating across the bridge. The next morning, American commanders peering through binoculars were stunned to see the bridge still intact, with fleeing German vehicles still retreating across. The local German commander had plenty of time to blow up the bridge, but still refused in order to let more of his troops escape to the east. American Lieutenant Karl Timmermann was ordered to seize the bridge. He faced stiff resistance from the last German's on the west side of the bridge. At 3:15 p.m. German engineers finally detonated a charge near the west side of the span, damaging it and making it temporarily impassable for tanks. Lt. Timmermann nevertheless sprinted across the bridge with his infantry. The Germans tried to blow up the central span, but the charges failed to detonate. Finally, another charge detonated and the bridge seemed to rise up in the air but then settled back down on its massive piers. In their haste, the German engineers had improperly placed detonators and the bridge had been built too well!

Courageous American infantry charged across the bridge, guns blazing, with hard fighting to follow. They then cleared the railroad tunnel (which the Germans should also have blown up) and secured the ridge overlooking the crossing. American combat engineers were able to make quick repairs to the damaged bridge, allowing troops and vehicles to cross. However, the bridge remained standing for only ten more days before collapsing from the pressures of heavy traffic and German attacks. It finally fell down altogether on March 17th.

Knowing the bridge was unstable, American engineers built pontoon bridges across the river nearby. Meanwhile, thousands of American troops, tanks, and heavy equipment were across the Rhine and poised to drive into the heart of Germany. The unexpected prize at Remagen forced the Allies to modify their strategy for invading Germany from Montgomery to the north.

Typical of American military doctrine, plans were modified on the fly as American commanders on the ground seized the initiative before them. This was in distinct contrast to Montgomery's plodding, ponderous methods of warfare. Some days would pass before Hodges broke out altogether from the new bridgehead. However, the US Army was across the Rhine.

Montgomery's Operation Plunder was scheduled to kick off on March 23rd and George Patton was well aware of that. Far south of Remagen and even farther from Montgomery, elements of Patton's Third Army had reached the west bank of the Rhine near the German city of Oppenheim the night of March 22nd. To his surprise, there was no welcoming party by the Germans on the east side of the river. Throughout that night, Patton ferried an entire division of American troops across the river on inflatable boats. That would be followed the next day by pontoon bridges erected by American combat engineers allowing tanks and heavy armor to cross. In the morning, Patton phoned Omar Bradley, his immediate superior. He said, "Brad, don't tell anyone, but I'm across." A surprised Bradley responded, "You mean across the Rhine?" "Sure am," Patton replied, "I sneaked a division over last night. But there are so few Krauts around there they don't know it yet. So don't make any announcement—we'll keep it a secret until we see how it goes."

By that night, the Germans indeed realized Patton had crossed the Rhine. But as far as Patton was concerned, an announcement was even more for his rival, Montgomery. Monty was preparing to cross the Rhine the next morning. So Patton called Bradley again. "Brad," he said, "tell the world we're across . . .. I want the world to know Third Army made it before Monty starts across."

The next day, Patton crossed the river on the pontoon bridge. When he reached the other side, Patton pretended to stumble, imitating William the Conqueror, who fell on his face when landing in England. William transformed his tumble into a positive by jumping up with a handful of English soil, proclaiming his coming conquest of England. Patton similarly arose, holding two handfuls of German soil in his hands and proclaimed, "Thus, William the Conqueror!"

On March 23, 1945, Eisenhower wrote a warm letter to Patton:

> "I have frequently had occasion to state, publicly, my appreciation of the great accomplishments of this Allied force during the past nine months. The purpose of this note is to express to you personally my deep appreciation of the splendid way in which you have conducted Third Army operations from the moment it entered battle last August 1. You have made your Army a fighting force that is not excelled in effectiveness by any other of equal size in the world, and I a.m. very proud of the fact that you, as one of the fighting commanders who has been with me from the beginning of the African campaign, have performed so brilliantly throughout. We are now fairly started on that phase of the campaign which I hope will be the final one. I know that Third Army will be in at the finish in the same decisive way that it has performed in all the preliminary battles.

Meanwhile, back at the northern end of the Rhine, Montgomery launched Operation Plunder on March 23rd. Monty's front stretched over 30 miles. During the night prior, British combat engineers constructed pontoon floats in preparation for the crossing of the Rhine in the morning. Bridge construction started at 9:45 a.m. and by 4:00 p.m. the first truck crossed the floating pontoon bridge. Over 1,152 feet of pontoon bridge was laid in the six hours and fifteen minutes. All in all, over 1.2 million British, Canadian, and American soldiers were ferried across the lower Rhine over the period of the next week.

Three Allied formations made the opening assault. Initially, there was little resistance, but later they ran into determined opposition from machine-gun nests. The US 30th Infantry Division landed south of Wesel, Germany. Local resistance had been largely eliminated by the artillery and air bombardments. Later, the US 79th Infantry Division also landed. American casualties were minimal. German resistance to the British sectors continued inflicting casualties and there were armored counterattacks as well. The landings continued, however, followed by tanks and other heavy equipment. By the evening of the 24th, US combat engineers opened another bridge across the river. As the assault pressed further eastward, they faced increasingly fierce German resistance. But by March 27th, the bridgehead was 35 miles wide and 20 miles deep into Germany.

Nazi propaganda minister Joseph Goebbels was well aware of Montgomery's assault and potential. On March 24th he wrote in his diary, "The situation in the West has entered an extraordinarily critical, ostensibly almost deadly, phase." He anticipated the crossings of the Rhine on a broad front and foresaw the Allies encircling the Ruhr industrial heartland. That is exactly what came to pass. It was the beginning of the end of the Third Reich.

CHAPTER 35

The Final Push

Throughout the war years, the leaders of England, America, and eventually Russia held high level conferences. In early February 1945, Roosevelt, Churchill, and Stalin—the Big Three—met at Yalta, in the Crimea of southern Russia. At other such conferences, there was discussion about the grand strategy for the coming year. But at the Yalta Conference, the discussion primarily was the plans for post-war Germany. It was agreed that each of the big three powers would occupy a zone of Germany. (The French quickly complained and were likewise allotted a zone as well.) The Soviet zone was the eastern portion of post-war Germany and Berlin was located in that zone. This would have a significant influence on the final battle plans of the Big Three vis à vis Germany.

British General Montgomery wanted to be the one to conquer Berlin. George Patton wanted to be the conquering hero of Berlin. And of course, the Russians wanted Berlin to themselves. But because Berlin would be in the Soviet zone of occupation, supreme commander of the western Allies, Dwight Eisenhower, ruled the Soviets would be allowed to take Berlin and the western Allies would stand down. This of course made unhappy campers of Patton and Monty. Nevertheless, Ike's order stood. His rationale in part was that American war planners estimated it would cost 100,000 American casualties to capture Berlin. As it turned out, the Red Army suffered over 360,000 casualties in sacking Berlin. Ike further reasoned, why shed American blood for what was largely a prestige achievement. This was particularly true when the United States would have to return the city to the Soviets in any event, as per the Yalta Agreement. In hindsight, Ike was right, though Churchill fretted about the growing Soviet menace after the war. To that degree, he was right as well. But this decision shaped the final battle plans of the war.

Massive Surrenders

Exploiting the seizure of the Ludendorff Bridge at Remagen on March 7, 1945, the US 12th Army Group under General Omar Bradley advanced rapidly into the German heartland. Their drive was to the *south* of German Field Marshal Model's Army Group B. To the *north*, the mostly British 21st Army Group under Montgomery crossed the Rhine on March 23rd. In a massive pincer movement, the forward elements of the two Allied army groups linked up on April 1, 1945. They were just to the east of the Ruhr region—the industrial heartland of Germany. In so doing, they created a massive encirclement of 317,000 German troops immediately to the west.

While the majority of Bradley's American divisions continued east towards the Elbe River, eighteen US divisions remained behind to destroy the entrapped German Army Group B. The noose was thus tightened around the German Ruhr pocket. For 13 days, the Germans held out against the American encirclement. But on April 14th German resistance crumbled. Model dissolved his army group on April 15th and ordered the Volksturm—old men recently forced into uniform—to discard their uniforms and go home. On April 16th the remaining German forces surrendered to the Americans. Model himself committed suicide on the afternoon of April 21st. The US Army suddenly had 317,000 German troops to secure and feed. They were placed in POW cages near Remagen.

Ike ordered Montgomery to then move northeastward with his British-Canadian forces basically along the German coast of the North Sea. He was to capture the important German naval-base cities of Bremerhaven and Hamburg—or what was left of them after years of Allied bombing. Though unstated, Ike also wanted Monty's forces to protect the northern American flank as they began to race into the heartland of Germany. Additionally, the Americans did not trust the Soviets and their expansionist intentions. Therefore, Monty's move across northern Germany prevented the Red Army from invading Denmark.

Meanwhile, American units began to race across Germany from Hodges First US Army in the center to Patton in the south. German forces initially fought with the frenzy of cornered beasts: some for reasons of Nazi fanaticism, some for patriotic reasons, and others to just survive. However, as the outcome of the war became increasingly clear, many German units just laid down their arms and surrendered when they encountered American or British forces. German soldiers to a man did not want to surrender to the Russians coming from the east. Therefore, many German units in the east made a hasty retreat westward, hoping to run into American forces and surrender to them. Literally tens of thousands of German soldiers simply laid down their arms and surrendered to the American onslaught. It was not uncommon to see a formation of thousands of surrendered German troops marching dejectedly to an Allied POW encampment (called cages during the war), guarded by just a handful of GIs marching beside them. An iconic picture taken during the war showed tens of thousands of surrendered German troops marching down the grass median of a German Autobahn (the fore-runner of the American interstate highway system) while an endless stream of Sherman tanks, US Army trucks full of soldiers, and other American heavy equipment moved down both sides of the divided highway.

Hitler, aware of all of this through reports reaching his underground bunker, issued orders for each German town and village to become a "Fortified Place." Accordingly, they were forbidden to surrender to the American onslaught. Local Nazi gauleiters were ordered to enforce his edict. Any sign of surrender or not supporting the Reich was to be dealt with summarily. Those suspected thereof were hanged from street lights or shot.

As American Army units approached a German town, if the phone system was working, a German speaking American solider might call the local phone operator and ask to be put through to the mayor or gauleiter of the town. When put through, the local official was told, "This is the US Army. We will approach your town in an hour (or whatever). If you surrender, we will not destroy your town. If you resist, your town will be turned to rubble." Some towns surrendered, some did not.

In one instance, a unit of Patton's Third Army approached a German town. An American lieutenant under a white truce flag was sent ahead with the ultimatum of surrender or be destroyed. A hot-headed Hitler Youth shot and killed the officer. Whereupon, Patton ordered the town reduced to rubble, which promptly took place.

In another instance, an American unit came upon an ancient German town which still had medieval walls, complete with a massive gate into the city. The town indicated they would not surrender. Whereupon the local American commander had a 155 mm Long-Tom heavy artillery piece towed forward. The massive gun was positioned, pointed toward the gate of the town. Then the artillery officer lowered the barrel of the gun to a basic horizontal position. He opened the breechblock of the gun, looked through the barrel at the target, made minor adjustments in aiming, loaded a shell, and fired. The gate of the city was obliterated. The German town promptly surrendered.

In other cases, white sheets would be hanging out the windows and from the poles of a town, indicating their willingness to surrender

On April 16th, American troops reached Nuremberg, the spiritual heart of the Third Reich where massive Nazi Party rallies were held each year. Here, Hitler had delivered some of his mesmerizing speeches. Hitler ordered the city held at all costs. Yet when 30 German soldiers approached the Americans carrying white flags, they were mowed down by Nazi machine guns to prevent their surrender. With German manpower virtually gone, the Nazis enlisted Hitler Youth to fight. Boys as young as 14-15 years old fought against battle-hardened US troops. The outcome was never in doubt. The city finally fell four days later. Ironically, that day, April 20, was Hitler's birthday.

On April 25, 1945, Soviet and American troops met and shook hands at the Elbe River, near Torgau in central Germany. Though only symbolic, it was an important step toward the end of World War II in Europe. The contact between the Red Army advancing from the East, and the US Army advancing from the West, meant the two powers had effectively cut Germany in half. It was not the end of the war, but it certainly portended it.

Meanwhile, Eisenhower ordered Patton to advance southeasterly. Rumors were rife that Hitler planned to leave Berlin and hole up in the Bavarian Alps. From there he could run a guerilla war against the Allies. It was called the National Redoubt. The plan was never fully endorsed by Hitler and no real attempt was made to implement the plan. However, it served as an effective propaganda tool as well as military deception by the Nazis in the final stages of the war. Notwithstanding, Eisenhower did not know that. He therefore ordered Patton, along with the American Seventh Army, southeastward to preclude any such operation. Upon finding nothing in that regard, Patton continued on toward Czechoslovakia and was in the region of Bavaria when the war ended. In the meanwhile, he cleared out virtually all Nazi opposition in southeastern Germany.

Horrific Discoveries

As the Allies pushed deeper into Germany, the true evil of Hitler and his Nazis became crystal clear, even through the fog of war. On April 4, 1945, the US Third Army came upon a group of large industrial-type buildings in the small town of Ohrdurf in south-central Germany. They quickly discovered this was part of a Nazi death camp complex. It was the first such camp encountered by American forces, and was part of the Buchenwald concentration camp network. (The Soviets had already discovered other camps in eastern Poland.) Inside this first camp liberated by US troops were dead bodies of victims starved to death and stacked like firewood. Close by was a pyre with charred skulls and bones left by Nazis guards. They fled upon seeing American troops approaching, knowing their work was an atrocity and had attempted to destroy evidence of their genocide.

Over a quarter-of-a-million prisoners, primarily Jews, had passed through the Buchenwald complex since it opened in 1937. The Nazis had built 110 primary concentration camps beginning in 1933, right after Hitler came to power. Six of them were major death or extermination camps. The Buchenwald complex was not designed so much as a mass killing camp, but rather for slave labor. As such, Jewish inmates died of starvation, disease, and abuse by the guards.

When Patton's army arrived, over 20,000 emaciated inmates were still there. They looked like ghosts. One witness said, "Their legs and arms were sticks with huge bulging joints, and their loins were fouled by their own excrement." On April 12th, a week after the camp's liberation, Generals Eisenhower, Patton, and Bradley toured the site, led by a prisoner familiar with the camp. Numerous corpses were found scattered around the camp grounds, lying where they were killed prior to the camp's evacuation. In a shed, a pile of 30 emaciated bodies were discovered, sprinkled with lime in an attempt to cover the smell. Patton, a man accustomed to the violence of war, refused to enter the shed as the sights and smells in the camp had already caused him to vomit against the side of a building. German citizens from the nearby town of Ohrdruf were forced to view the camp and bury the dead, a practice that was later repeated in other camp liberations. Following the tour, the mayor of Ohrdruf and his wife went home and hung themselves.

The British Army found 60,000 emaciated inmates at Bergen-Belson jammed into a camp designed to hold about 2,000. No real attempt had been made to feed these poor souls who were wasting away from starvation. There was no sanitation. The floors of the inmates' huts were covered with human feces and stale urine. As a result typhus was raging through the camp. One British officer wrote that the inmates there were little more than "living skeletons with haggard, yellow faces." Many were on the edge of death. Five-hundred people a day died. There were dead men and women, all naked, lying in heaps on both sides of a railroad track. Those still alive walked slowly and aimlessly with a vacant expression on their famished faces.

Richard Dimbleby, a BBC reporter accompanying the British troops who came upon Bergen-Belson sent a communique back to London. Below is an edited version of his account.

> "Beyond the barrier was a swirling cloud of dust .. .. And with the dust was a smell sickly and thick of death and decay . . .. I passed through the barrier and found myself in the world of nightmare. Dead bodies . . . lay strewn about the road and along the rutted track. On each side of the road were brown wooden huts. There were faces at windows, the bony emaciated faces of starving women, too weak to come outside.
>
> I have seen many terrible sights in the last five years, but nothing, nothing approaching the dreadful interior of this hut in Belsen . . .. The dead and the dying lay close together. I picked my way over corpse after corpse . . . until I heard one voice that rose above the gentle undulating moaning. I found a girl. She was a living skeleton. Impossible to gauge her age because she had practically no hair left on her head and her face was only a yellow parchment sheet with two holes in it for eyes . . . beyond her . . . there were the convulsive movement of dying people too weak to raise themselves from the floor.
>
> One woman, distraught to the point of madness, flung herself at a British soldier . . . she begged him to give her some milk for the tiny baby she

held in her arms . . . she put the baby in his arms
and ran off crying . . . when the soldier opened the
bundle of rags to look at the child he found it had
been dead for days.

Women stood naked at the side of the track,
washing in cupfuls of water taken from the British
Army water trucks. Others squatted while they
searched themselves for lice . . . helplessly, and all
around was this awful drifting tide of exhausted
people neither caring nor watching.

The Brits had gotten a vision of the hell created by the Fuhrer of
Germany, Adolf Hitler. Not only was he an immoral dummkopf,
he was evil incarnate—empowered by the Devil. The concentration camps of Germany were a vivid testament of a man possessed by Satan. The Nazis did not care if the inmates lived or
died. Along with the more than one hundred concentration
camps, the six death camps were designed specifically for assembly-line death operations. If there were anything good about
these hell holes, the victims in the death camps often were led
from the box cars they had been packed into directly into the
gas chambers, not suffering the terrible lingering deaths en-
dured in the work camps. Such was the offspring of the wicked,
warped ideology and philosophy of Hitler and the Nazis. Ideas
have consequences and the Nazi concentration camps were the
visible manifestation of Nazi ideology. As the Allies swept into
Germany from west and east, the ghastly concentration camps
were discovered one by one.

It is evident the Nazis knew their depravity and vainly tried to
hide their monstrosities. However, as Allied troops rapidly swept
across Germany in the final month of the war, the Nazis could not
cover the atrocities. At Auschwitz, over 2,000 prisoners were still

alive when the camp was liberated and lived to tell their stories. The Allies found warehouses full of clothes removed from those headed to the death chambers. Even more grotesque were bales of human hair the Nazi intended to use for industrial purposes. Rooms full of eyeglasses and artificial limbs were a testimony to the depravity of their captors. The Nazis routinely looted gold and silver dental fillings from their victims' teeth to enrich their coffers. The abominations were often beyond description. Many books have been written detailing the Nazi atrocities. We can only give a brief summary here.

Dwight Eisenhower would write upon viewing the ghastly camp at Ohrdruf, "We are told the American soldier does not know what he is fighting for. Now, at least, we know what he is fighting against."

CHAPTER 36

The Battle for Berlin

During January and February 1945, the Red Army advanced through what had formerly been Poland, moving up to the Oder River. Portions of the Oder had been roughly the border between Germany and Poland before the war. In a little over two weeks, the Red Army had advanced 300 miles from the Vistula River to the Oder. They were only 43 miles from Berlin. However, Russian Field Marshall Zhukov called a halt to neutralize opposition on his northern flank in the Pomerania region. During the month of March, they rested, refitted, and prepared for the upcoming battle. The final drive on Berlin would not commence until April.

The Soviets resumed their offensive on April 16th. The Russian high command planned an encirclement of Berlin from the east, southwest, and north. Soviet Marshall Zhukov commanded the Second Belorussian Army group which would attack from the east. Marshall Ivan Konev commanded the First Ukrainian Army group and was to swing to the south of Berlin and attack from the southwest. Both men were rivals. Stalin pitted Zhukov against Konev by challenging them to see which could reach Berlin first. Zhukov had the shorter distance to Berlin but faced the heaviest resistance. Konev had a longer distance to go by making an end-run around the south of Berlin and then attacking to the northeast. However, he would face less German defenses. Meanwhile to the north, another Russian army was poised to overcome German defenses north of Berlin.

Prior to the war, Berlin had been one of the largest cities in the world with a population around four-and-one-half million. Before the main fighting commenced, the Red Army surrounded the city after successful battles in its eastern suburbs. On April 20, 1945, Hitler's birthday, the First Belorussian Front led by Zhukov, started shelling central Berlin. Meanwhile Konev's First Ukrainian Front broke through German defenses and advanced into the southern suburbs of Berlin.

On April 23rd, Hitler appointed General Weidling as commander of all German forces within Berlin. At his disposal were several depleted and disorganized Wehrmacht and Waffen-SS divisions. He also had poorly trained Volksturm (older men forced into service) along with fanatical Hitler Youth members. Over the course of the next week, the Red Army gradually took the entire city. All in all, it took the Red Army a little over two weeks to conquer Berlin.

As the Allies, both east and west, closed in on Germany, remaining Wehrmacht troops, not able to surrender or not willing, fell back toward Berlin. Their remnants became concentrated around the city. They in fact became like a wild animal which had been cornered—they fought to the death. The fighting against the Russians was therefore fierce and intense, but the battle was hopelessly lopsided. The Red Army fielded about two-and-one-half million troops against Berlin compared to perhaps 750,000 German soldiers in the vicinity. The Russians had over 41,000 artillery pieces compared to 9,300 for the Germans. Once within the city limits, the Russians still had about 1.5 million troops against 45,000 Wehrmacht and SS troops and another 40,000 Volksturm.

The Red Army shelled the center of Berlin incessantly where the government buildings were located. What American and British bombing had not destroyed over the preceding several years, Russian artillery nearly finished the job. The battle eventually devolved into street fighting with small battles raging from house to house and door to door.

In the wee hours of April 29th, the Soviet Third Shock Army crossed into the area of the Reichstag and began fanning out into the surrounding streets and buildings. The initial assaults on buildings, including the Ministry of the Interior and Gestapo Headquarters were hampered by the lack of artillery support caused by knocked out bridges. By the next day, April 30th, the Russians had solved the bridge problems and now had artillery support. At 6 a.m. they launched an attack on the Reichstag itself. However, because of German resistance, it was not until that evening that Russian soldiers were able to enter the building. The Reichstag had not been used by the Germans since it burned in February 1933. The interior was essentially a heap of rubble. The Germans inside thus made excellent use of it and fortified themselves in the wreckage. The result was intense room-to-room fighting. However, by May 2nd, the Red Army controlled the building entirely.

The seizing of the Reichstag in Berlin became a symbol to the Russians of not only victory in Berlin, but of the entire battle with Nazi Germany. Thereafter, a Soviet commander ordered, "The Supreme High Command . . . and the entire Soviet People order you to erect the victory banner on the roof above Berlin." The famous photo of the Russian soldiers planting the hammer-and-cycle flag on the roof of the building was actually a reenactment which was taken the next day.

Yet, the battle was not quite over. As the German defensive perimeter shrank and surviving German defenders fell back, they became concentrated into a small area in the center of the city. Ten thousand German soldiers were there and were being assaulted on all sides. The Red Army effectively cut off the last area held by the Germans making any escape attempt to the west virtually impossible.

Early in the morning of May 2nd, as the Soviets captured the Reich Chancellery, the final German commander in Berlin, General Weidling, surrendered with his staff at 6:00 a.m. He was taken to General Vasily Chuikov where Weidling ordered the city's defenders to surrender to the Soviets. Meanwhile, on the night of May 2nd, General von Manteuffel, commander of the Third Panzer Army along with General von Tippelskirch, commander of the XXI Army, surrendered to the US Army.

Though victorious, the Russians paid a heavy price for their prize. The Nazi beast did not go quietly but fought back viciously to the very end. Russian records show that 81,116 Russian soldiers were killed in the Battle for Berlin and another 280,251 were wounded. The Germans suffered around 100,000 deaths, 220,000 wounded and 480,000 taken prisoner. An estimated 120,000 civilians also died in the overall battle of Berlin, including its suburbs.

But the suffering of Nazi Berlin was not over. The Russians never forgot how the Wehrmacht and SS units committed atrocities when they invaded Russia four years earlier. Russian villages were burned. Peasants had all their belongings stolen and if they protested were shot on the spot. Men, women, and children were massacred. And, the average Russian was also aware of how the SS had murdered every Russian Jew they could find. It was another of Hitler's blunders. Russian peasants had no love for Stalin and if Hitler had treated them decently, they would have joined with the German army in overthrowing Stalin. To the contrary, Hitler turned them into bitter enemies.

When the Red Army moved into Germany, Russian troops were not only seeking victory, they were seeking revenge. This reached epidemic proportions as they conquered Berlin. Wehrmacht troops had no illusions of the tender mercies of the Russians toward them. Many who were captured died of starvation in Russian POW camps. The Russian Army either did not have food to feed them, or did not care—probably both.

But German women suffered even more. As fighting settled down at night, Russian troops went searching basements where civilians were hiding and grabbed the women and girls. In broken German they ordered the equivalent of "lay frau lay," whereupon they raped them. Initially, Soviet commanders allowed and even encouraged such behavior. Eventually, Red Army authorities cracked down and instilled military discipline on their troops in this regard. Nevertheless, in the occupation after the fall of Germany, the raping continued. Estimates range as high as two million German women and girls having been raped by Russian soldiers. Many were raped multiple times. Not a few committed suicide thereafter. For many a German female, their fate after the war was one worse than death. For some Russian soldiers, the base male impulse for sexual intercourse may have been the motivation; but for most, it was more out of revenge. When one Soviet officer was rebuked for the behavior of his men after the war, he replied, "Russian soldiers do not kill women" (an allusion to the atrocities of the Wehrmacht and SS). Unspoken, but clearly implied was, "but we might rape them."

After the fighting died down, the Russians did attempt to restore some basic services. Soviet authorities took measures to start restoring essential utilities. Almost all transportation in and out of the city had been wrecked. Bombed-out sewers had contaminated the city's water supplies. The Red Army appointed local Germans to head each city block and begin organizing cleanup. The Russians also made some efforts to feed the residents of the city with soup kitchens, but for most citizens of the Berlin area, starvation hovered. Germany and its major cities, particularly in the east, faced terrible times after the war. Germany had not only been defeated, it had been destroyed.

CHAPTER 37

Hitler's Last Days

The last four months of Hitler's life was nothing but one report after another of bad news. By early 1945, Nazi Germany was on the verge of total collapse militarily. The Russians had advanced through Poland and by April were only 40 some miles from Berlin. In the west, the Americans had turned the battle of the Bulge into a defeat for Hitler and a victory for the US Army. The Rhine had been crossed, the Ruhr had fallen along with the capture of 317,000 German troops. As the Russian bear bore down relentlessly from the east, the Americans were racing through the heartland of Germany and had crossed the Elbe River in central Germany.

On January 16, 1945, Hitler moved into the recently rebuilt Fuhrer Bunker located just outside the Reich Chancellery in downtown Berlin. The bunker had originally been built in 1936 as an air raid shelter adjacent to the Chancellery. In 1944, it was expanded into two levels, the deepest being 28 feet below ground. The lower, newer level would become Hitler's home and headquarters for the last four months of his life. The bunker occupied approximately 3,000 square feet, was self-sufficient, and considered impregnable. It had its own electric generation plant, well water, air conditioning, and heat as well as a ventilation system. The bunker was covered with approximately ten feet of reinforced concrete. Though all concrete in its construction, it had been decorated with high-quality furniture taken from the Chancellery, along with several framed oil paintings. There were bedrooms for Hitler, his mistress Eva Braun, offices, a conference room, quarters for the Goebbels family, along with various other miscellaneous rooms for utilitarian purposes.

By April 19, 1945, what was left of the Wehrmacht was in full retreat westward from Seelow Heights, an eastern suburb of Berlin, the last defensive line of the city to the east—the Russians in hot pursuit. On April 20th, Hitler's birthday, the Red Army began artillery bombardment of the city. By the evening of April 21st, Russian tanks had reached the outskirts of Berlin. On April 22nd, Hitler learned orders he had issued the previous day for SS General Steiner to counterattack had not been obeyed. He collapsed in state of total nervous exhaustion.

As the news got only worse for him, Hitler began to drift into delusion. He had always been a desk commander—that is, he laid out maps, studied them, and then gave orders based upon the situation he saw on his maps, even though the battles might be raging hundreds of mile away. He had long ago lost any sense of what was really going on in the various battles he tried to micromanage. Now as the end was near, he imagined German army divisions which did not exist coming to the rescue to deliver Berlin. He entertained delusional thoughts that somehow fate would look kindly upon him and grant some miraculous victory such as had happened to Frederick the Great in the 18th century.

Hitler's health had been deteriorating for years. He had entrusted himself to his quack physician, Dr. Theodor Morell, who supplied him with various concoctions of unorthodox drugs and even narcotics. Hitler had become haggard and stooped. His countenance was pale and washed out. He trembled and very likely had Parkinson's disease. He looked years older than he was. Such was the supreme leader of the alleged Aryan super race. At 56 years of age, he was a physical basket case and by late April 1945, he had become a nervous wreck.

Upon hearing the news of Steiner's insubordination on April 22nd, Hitler launched into a tantrum against the commanders present before him, calling them treacherous and incompetent, declaring to them—for the first time—the war was lost. Hitler announced to all present he planned to remain in Berlin until the bitter end and then shoot himself. When Luftwaffe chief and Reichsmarshall Hermann Goering learned of this, he sent a telegram to Hitler requesting permission to become the leader of the Reich as Hitler had promised in 1941, naming him as his successor. Hitler therefore informed Goering he would be executed unless he resigned all of his posts. Later that day, he relieved Goering from all of his leadership and ordered his arrest.

By April 27th, the Russian encirclement of Berlin left it cut off from the rest of Germany. Hitler no longer had secure radio communications with the last remnants of his armies real and imagined. His staff in the Fuhrer Bunker had to rely upon what was left of public telephone lines for issuing orders. The only news Hitler got was from public radio news bulletins. On April 28th, Hitler heard a BBC report which said Heinrich Himmler had offered to surrender to the Western Allies. The offer was declined. Himmler implied he had the authority to negotiate a surrender, which Hitler considered treason. Later that day, Hitler exploded and ordered Himmler arrested.

Also on April 28th, Hitler learned the Red Army had advanced into the Potsdamer Plaza in central Berlin, only a mile and a half from the Reich Chancellery and his bunker beneath. It was apparent they planned to assault the Chancellery the next day. Late that night, not long after midnight (now April 29th), Hitler married his long-time mistress Eva Braun in a small civil ceremony in the map room of the Fuhrer Bunker. That morning, the Hitlers hosted a modest wedding breakfast. He then took his secretary Traudl Junge to another room and dictated his last will and testament.

Hitler's last testament was a rambling rehearsal of the failure of his subordinates. Among other things, he put into writing his wishes for the Reich government and who was to succeed him after his death. Notable was that Admiral Karl Doenitz was to be the new head of state. Goebbels was to become the new chancellor of Germany.

In his testament, Hitler stated several times that he never wanted to go to war and wished for only peace. But laced throughout it was repeated assignation of blame on the Jews. He wrote:

> "Centuries may pass, but out of the ruins of our cities and monuments of art there will arise anew the hatred for the people who alone are ultimately responsible: International Jewry and its helpers!"

> "As late as three days before the outbreak of the German-Polish War, I proposed to the British Ambassador in Berlin a solution for the German-Polish problem—similar to the problem of the Saar area, under international control. This offer cannot be explained away, either. It was only rejected because the responsible circles in English politics wanted the war, partly in the expectation of business advantages, partly driven by propaganda promoted by international Jewry."

"After six years of struggle, which in spite of all reverses will go down in history as the most glorious and most courageous manifestation of a people's will to live. I cannot separate myself from the city which is the capital of this Reich. Because our forces are too few to permit any further resistance against the enemy's assaults, and because individual resistance is rendered valueless by blinded and characterless scoundrels, I desire to share the fate that millions of others have taken upon themselves, in that I shall remain in this city. Furthermore, I do not want to fall into the hands of enemies who for the delectation of the hate-riddled masses require a new spectacle promoted <u>by the Jews</u>."

"Above all, I charge the leadership of the nation and their followers with the strict observance of the racial laws and with merciless resistance against the universal poisoners of all peoples, <u>international Jewry</u>."

Given in Berlin, 29th April 1945, 4:00 a.m.
Signed: A. Hitler

His testament is rife with the views of a man who was either out of touch with reality or was a pathological liar. To the very end, he was obsessed with blaming all the troubles of the world, and his in particular, on the Jews. He was a man who at the end had become deranged and delusional in his hatred for the sons of Israel. Like many other haters and oppressors of the Jews down through the centuries, he came to an ignominious end. God foretold to Israel thousands of years earlier, "I will bless them that bless thee, and curse them that curse thee." God's curse upon

Hitler was manifest in a broad spectrum. But one thing is for sure, he died a miserable death as a miserable man and will be for all eternity.

Later that day, April 29th, Hitler learned his ally, Benito Mussolini, had been executed by Italian partisans. His body along with that of his mistress, Clara Petacci, had been strung up by their ankles. Their corpses were later cut down and thrown into the gutter, where they were mocked by ordinary Italians. These events very well may have influenced Hitler not to allow himself or his new wife to be made a "spectacle" as he had earlier worried about in his testament.

At one in the morning on April 30th, Field Marshal Keitel reported that all the forces upon which Hitler was depending to rescue Berlin had either been encircled or no longer existed. At around 2:30 that morning, Hitler met with about twenty people, mostly women, who had gathered to say farewell. He walked among them and shook hands with each of them and then went to bed. Upon rising late that morning, the Red Army was about 500 yards from the Fuhrer Bunker. General Weidling, the commander of the Berlin Defense Area therefore met with Hitler. Weidling told Hitler that German defenders in the city would probably run out of ammunition that night. He also ominously told Hitler the fighting in Berlin would more than likely end within the next 24 hours.

Hitler, his two secretaries, and his personal cook then had lunch. Then, he and his wife said goodbye to his bunker staff and fellow occupants, including Bormann, Goebbels, the secretaries, and several military officers. At around 2:30 that afternoon, Adolf and Eva Hitler went into his personal study. Witnesses later reported that they heard a loud gunshot at about 3:30 p.m. Aides shortly thereafter entered and immediately noted a scent of burnt almonds, the tell-tale odor of prussic acid (hydrogen cyanide). Inside the study they found the two lifeless bodies on the sofa. Eva, with her legs drawn up was to Hitler's left and slumped away from him. He was slumped over with blood dripping out of his right temple. He had shot himself with his own pistol. His head was lying on the table in front of him. Eva's body had no visible physical wounds, but her face showed how she had died—by cyanide poisoning. In accordance with Hitler's prior written and verbal instructions, the two bodies were carried up the stairs and out the bunker's emergency exit to the garden behind the Reich Chancellery. There they were to be burned with gasoline. The funeral pyre did not burn well at the first. Finally, SS men brought additional cans of gasoline and with the ensuing fire completely incinerated Hitler and his wife. Their remains burned until 6:30 that evening whereupon the remains were buried in a shallow grave in a nearby bomb crater.

Thus ended the life of one of the most evil characters in history. His name to this day is a name of ignominy.

German Surrender

With Hitler dead, Berlin in Russian hands, and most front-line German armies defeated or languishing in prisoner-of-war cages, the end was evident. Hitler's designated successor, Admiral Karl Doenitz, was the new titular head of Germany. During the month of April 1945, the total number of prisoners taken by the Western Allies alone was 1,500,000. Over 800,000 German soldiers had surrendered or been captured by the Russians since the beginning of 1945.

Little by little, German forces in outlying areas began to lay down their arms. In Italy, the German army surrendered to the Allies on April 29, 1945. German forces in Northwest Germany, Denmark, and the Netherlands surrendered May 4th. German forces in Bavaria surrender on May 5th. Nazi forces in Norway surrendered on May 7th.

Eisenhower's Supreme Headquarters Allied Expeditionary Force (SHAEF) was located in a school house in the town of Reims in northeastern France. On May 6th, the Chief-of-Staff of the German Armed Forces High Command, General Jodl, arrived in Reims and following Doenitz's instructions, offered to surrender all forces fighting the *Western* Allies. Doenitz undoubtedly knew Eisenhower would reject such an offer. The position of the Allies always had been *unconditional* surrender—to all of them, including the Russians. But Doenitz was trying to buy time for Wehrmacht troops in the east to manage to escape westward and surrender to either the British or the Americans. Eisenhower threatened to walk away unless the Germans agreed to a total and unconditional surrender to all the Allies. Ike further told Jodl unless Doenitz accepted his terms, he would order western lines closed to German soldiers, forcing them to surrender to the Soviets. Jodl quickly sent a message to Doenitz, in Flensburg, Germany informing him of Eisenhower's statement. Shortly after midnight on May 7th, Doenitz, accepted Ike's ultimatum. He sent a radio message to Jodl authorizing him to sign a complete and unconditional surrender of all German forces.

At 2:41 on the morning of May 7th, at SHAEF headquarters, the Chief-of-Staff of the German Armed Forces High Command, General Alfred Jodl, signed an unconditional surrender document for all German forces to the Allies. The instrument of surrender included the phrase, "All forces under German control to cease active operations at 2301 hours Central European Time on May 8, 1945."

The next day, May 9th, Field Marshal Wilhelm Keitel and other high ranking German representatives traveled to Berlin, and shortly before midnight signed another document of unconditional surrender, again surrendering to all the Allied forces, this time in the presence of Marshal Georgi Zhukov of the Soviet Union. Stalin had insisted on the latter.

The war in Europe was over.

The Aftermath

Many books have been written about the aftermath of World War II. It is not our purpose to examine all the consequences of the war—there are many. But there were major changes across the world in the years immediately after World War II. Germany was occupied and divided into four zones: the American, Soviet, British, and French. Each of these major powers were also allotted a zone of occupation in Berlin. Following the war came the Nuremberg war crimes trials. Notable Nazis were tried; some were hanged, others imprisoned, and a few exonerated. In June 1948, the Soviets attempted to push the other Allies out of Berlin and sealed off the city. The only thing preventing starvation and hypothermia that winter was a major effort led by the United States to airlift food, coal, and necessary supplies to West Berlin. It was called the Berlin Airlift and lasted almost a year. Also in 1948, the United States implemented the Marshall Plan to aid war-ravaged Europe. Western Europe, though on the winning side, had had its economies and farms largely ruined by the war. The United States therefore sent massive amounts of aid, both cash and food, to prevent Europe from starving and being overrun by the insurgent communist movement, especially virulent in troubled countries. The Soviet Union particularly was behind the communist expansion around the world, especially in Europe. As a result, the North Atlantic Treaty Organization (NATO) was formed in 1949 to present a united front against the Soviets. In turn, the Cold War descended upon the western world and would continue for another 40 or so years. Notwithstanding, there were no *major* shooting wars during this period.

CHAPTER 39

The Rebirth of Israel

A nother consequential event took place in the years immediately after the war—the rebirth of the State of Israel. Israel as a sovereign state ceased to exist in about 605 B.C. Israelites were removed from Israel by the Assyrians in 721 B.C. and then by the Babylonians beginning in 605 B.C. Jews were allowed to return to their land by the Persians beginning in about 535 B.C., though many had become comfortable and affluent living in gentile countries of the day. By the time of Christ, however, there was a quasi-de facto nation of Jews—Judaea and Galilee—living in what is today called Israel. However, they had been under the suzerainty of first Persia, then Greece, and finally Rome in the decades prior to and then after the life of Christ. Rome in fact occupied that region in the first century B.C. In A.D. 70, Rome crushed a Jewish revolt and again in A.D. 136. As a consequence, they deported all Jews out of "the land." It is known by Jews to this day as the *diaspora*— the dispersion. They eventually came to live all across Europe, and around the world. By the nineteenth and early twentieth centuries, Jews were an integral part of Europe and particularly Eastern Europe. Nevertheless, for centuries they had been persecuted and forced to move often. Nations dominated by the Roman Catholic Church, the several Orthodox bodies, and even the major Protestant denominations viewed them as responsible for the crucifixion of Jesus Christ. Jews were routinely referred to as "Christ killers." In turn, they were often harassed and persecuted. The Tsars of Russia regularly enacted *pogroms* (organized, government-sanctioned persecution and even massacres).

As noted throughout this book, Hitler and the Nazis came to blame the Jews for all the evil, wars, and difficulties visited upon the world and upon Germany in particular. In turn, he became obsessed with exterminating the Jewish race insofar as he was able to do so. The Holocaust was the result. However, what is utterly ironic is that the blood of the Holocaust became the seed of the rebirth of the State of Israel. Hitler had attempted to stamp out Jewry across Europe, but the State of Israel was re-constituted on May 14, 1948. It is directly attributable to the Holocaust.

In the years prior to World War II, there were about 15 million Jews living around the world. Hitler managed to murder about six million of them in his abominable death camps and other Nazi murder programs. But after the war, for the Jews remaining in Europe and other nations, a yearning erupted to return to "the land"—the land which God had promised Abraham, Isaac, and Israel.

In the late 19th century, an Austro-Hungarian Jew by the name of Theodore Herzl formed the Zionist Organization and promoted Jewish immigration to Palestine in an effort to form an independent Jewish state. That movement to this day is called Zionism—Mount Zion being a principal hill or low mountain located in Jerusalem. From about A.D. 1299 until the end of World War I, the region commonly called Palestine, Lebanon, and Syria, among others, had been ruled by the old Turkish Ottoman Empire. Turkey wound up on the losing side of World War I. As a result, Lord Balfour, one-time Prime Minister of England and Foreign Secretary of the United Kingdom at the end of World War I, issued a proclamation ever since known as the Balfour Declaration. As one of the victors, the British government in 1917 thus issued the Balfour Declaration at the end of World War I declaring support for the establishment of a national home for the Jewish people—in Palestine.

The name *Palestine* was coined by the Romans during their occupation of the region and used particularly after the Jews were scattered out of the land in A.D. 70 and 136. It was the Latinized pronunciation of *Philistina* or the land of the ancient Philistines— Greek-Phoenician marauders who settled coastal areas of what later would be called Palestine. Rome wanted to erase all connection to the region of the name Judaea or Israel and therefore came up with the substitute term *Palestina*.

Because of the Balfour Declaration, Jews began to trickle back to what they have called for millennia as "the land" (*ha eretz* in Hebrew). The latter refers to how God promised to Abraham "the land" in the Book of Genesis almost 4,000 years ago. In the early twentieth century "the land" was desert, swamps, and sparsely populated. As Jews began trickling back from the diaspora, they purchased land from the small Arab population and began to build towns—Tel Aviv, a notable example. From 1922 onward, the British administered the region in what was called the Palestinian Mandate, granted by the League Nations after World War I. However, in the ensuing several decades, there was substantial friction between the local Arabs, arriving Jews, and the British authorities.

As the Nazis came to power in Germany, the Arab-Islamic leaders in Jerusalem, the Mufti of Jerusalem in particular, openly supported Hitler and the Holocaust. This only served to further exacerbate friction between the Jews in the land and the small local Arab population.

After World War II ended, Zionism became more than a theoretical goal for most Jews living in Europe. They had had enough. In the first part of the twentieth century, many Jews living in Europe were comfortable, established in their communities, and many even affluent. They had little motivation or interest in uprooting and moving to primitive conditions in Palestine. But the war and the Holocaust changed all of that. Surviving Jews, many of whom were displaced persons and having nowhere else to go, determined to return to "the land" of ancient Israel. They began to flood into Israel by any means possible.

The British authorities of Mandatory Palestine eventually tried to stop the "exodus" of Jews arriving from elsewhere. In fact, a Jewish organization bought an old ship and named it the "Exodus." Four-and-one-half thousand Jews, mostly survivors of the Holocaust, crowded aboard the ship from a port in southern France. It was quickly intercepted by the British Royal Navy and violently harassed as it neared Haifa. The British opened fire and forcibly boarded the ship killing Jewish refugees and wounding others. The British, to their shame, turned the ship around and the Jewish refugees on board wound up back in Germany. Britain by then was motivated by an interest in Arab oil and maintaining what was left of its tattered empire in the Middle East. Their policies therefore came to favor the Arabs over the Jews.

Meanwhile, Jewish resistance organizations sprang up in "the land." One such group, the Haganah, was a mainstream paramilitary organization which fought against not only the Arabs before and after World War II, but particularly against the British afterwards. The Haganah would eventually become the fledgling Israel Defense Force (IDF) after the rebirth of Israel. However, there were several more radical paramilitary groups operating, one of which was the Irgun. Whereas the Haganah operated primarily as saboteurs of British installations, the Irgun was not reluctant to kill Arab foes but the British as well. In the years 1946-1948, the British were in endless conflict trying to suppress the several Israeli paramilitary groups as well as pacify the Arabs. Things came to a head in July of 1946 when the Irgun blew up the King David Hotel in downtown Jerusalem. The hotel was the British administrative headquarters for Palestine. In the explosion, 91 people of various nationalities were killed, and 46 were injured.

Britain was exhausted from World War II. They perceived no benefit from Palestine. Much political maneuvering, including the newly-formed United Nations, went on for several years. But the Brits had had it. They therefore determined to depart. On April 29, 1948, England announced British forces would be withdrawn from Palestine on May 14, 1948. That they did and that day, May 14, 1948, the Jews living in Palestine declared themselves an independent nation. The infant State of Israel immediately found itself at war with the surrounding Arab world, led by Jordan, Egypt and the Arab League. Tiny Israel with a population of about 600,000 Jews was attacked by Arab nations with a combined population of 110 million. Yet as with David and Goliath, tiny Israel prevailed.

In the greater scope of the aftermath of World War II, one of the major and long-lasting results was the rebirth of Israel. It undoubtedly was a fulfillment of biblical prophecy foretold thousands of years ago of the re-gathering and return of Jews to their land. That prophecy has not as yet been completely fulfilled—millions of Jews still are scattered around the world, but the fulfillment has begun. In hindsight, a conclusion might be made that God allowed the Holocaust as a catalyst to provoke Jews to return to the land of their heritage. The irony of it all is that in trying to eradicate the Jewish race, Hitler was instrumental in their re-gathering and the formation of Israel as a sovereign state, with a place among the nations.

World War II was the greatest secular event in human history. A misfit of mediocre background, within several years, rose to be the most powerful individual in the world. In the view of this author, Hitler's contact with the occult led him to be possessed with satanic power. There is no other explanation for his meteoric rise to power and his categorical, maniacal hatred of the Jews. Though being energized by the Devil, he made one blunder after another which led to his ultimate destruction. The name Adolf Hitler will go down in history as not only one whose pride, hatred, and arrogance led him to make blunder after blunder, ensuring the defeat of Nazi Germany, but as evil incarnate.

* * * * *

www.ingramcontent.com/pod-product-compliance
Lightning Source LLC
Chambersburg PA
CBHW022053210326
41519CB00054B/325